ArcPy and ArcGIS

Second Edition

Automating ArcGIS for Desktop and ArcGIS Online with Python

Silas Toms
Dara O'Beirne

BIRMINGHAM - MUMBAI

ArcPy and ArcGIS

Second Edition

First published: Feb 2015

Second edition: June 2017

Production reference: 1270617

Published by Packt Publishing Ltd.
Livery Place
35 Livery Street
Birmingham
B3 2PB, UK.

ISBN 978-1-78728-251-3

www.packtpub.com

Credits

Authors
Silas Toms
Dara O'Beirne

Reviewers
Ken Doman
Adeyemo Ayodele Oba

Commissioning Editor
Aaron Lazar

Acquisition Editor
Karan Sadawana

Content Development Editor
Kinnari Sanghvi

Technical Editor
Kunal Mali

Copy Editor
Safis Editing

Project Coordinator
Vaidehi Sawant

Proofreader
Safis Editing

Indexer
Francy Puthiry

Graphics
Abhinash Sahu

Production Coordinator
Aparna Bhagat

Foreword

I have long preached the importance of GIS programming, and students who have listened have done well and not spent all of their time clicking their way through their work. They've also gotten good jobs and interesting careers. However, learning GIS programming needs resources, and while there are a lot of clues on websites and you can certainly learn programming languages such as Python in places like CodeAcademy, there's a *lot* more to GIS programming, and understanding some of the complexities of accessing spatial data and processing requires a systematic approach. This new edition of *ArcPy and ArcGIS: Geospatial Analysis with Python* by Silas Toms and Dara O'Beirne is one such approach and it provides a significant contribution, extending from desktop to online GIS methods.

Toms and O'Beirne take the approach of jumping right into the deep end and exploring the water, or as I'd prefer and the original cover illustration suggests, jumping in and exploring a mountain river. This approach is particularly useful for those of us who already have some experience with programming and just need some clues about how to work in this particular pool and find its eddies and falls. You can spend an entire book learning about the Python language, and while that can be valuable, who has the time?

There's a lot to the ArcPy part of Python scripting; a lot more than the language itself. ArcPy scripting is very much a high-level scripting approach, but the stuff you interact with has considerable depth. You need to understand how to use the geoprocessing tools, ArcPy methods and properties, spatial data relationships, and many other structures. So reasonably, this book focuses on the ArcPy part and takes the user all the way to online GIS automation, which depends more on knowing how to find and structure things than knowing how to code.

Learning about cursors in Chapter 3, *ArcPy Cursors - Search, Insert, and Update,* and geometries in Chapter 4, *ArcPy Geometry Objects and Cursors,* gets right to the heart of the value of Python scripting in ArcGIS. The most useful ArcPy scripts I've ever used or written have used cursors, mostly accessing and creating geometries. This is where you really see the benefit of a scripting model, and you can write tools that can do things that are impossible (or at least impractical; I guess you can edit each feature one at a time) without a script.

Continuing in the deep river pool mode—gets your heart racing—we start building script tools in `Chapter 5`, *Creating a Script Tool*! Now when I write scripts, I tend to have to fight my way through bugs—I'm a bit of an empirical programmer—so I tend to first get my code working with hardcoded data in the IDE at the script tool stage. At the script tool stage, `arcpy.AddMessage` becomes my typical debugging tool, serving as the venerable print statement, something like `arcpy.AddMessage` ("Just finished the data conversion loop"). However, while many readers may want to follow my approach, building a working script tool in `Chapter 5`, *Creating a Script Tool*, lets you see where you can get pretty quickly.

Many GIS users are working in agencies dealing with managing lands or infrastructure, and with this type of responsibility, building maps is a foremost concern. The chapter on the `Arcpy.mapping` module is especially welcome, because we don't just need to automate our analysis, we also need to automate our maps for greater efficiency and consistency in the design.

The direction of GIS these days is certainly online, so the significant coverage of online methods in this edition, ranging from ArcGIS Online and REST services to the new ArcGIS API for Python, is welcome. While I've traditionally focused on desktop analytical methods, I plan to use this introduction to move some of our toolsets online. Finally, even we die-hard desktop users have learned that we need to move online at least enough to make good use of ArcGIS Pro, so this coverage is a welcome addition.

All in all, Toms and O'Beirne have certainly come up with a very useful text that certainly belongs on the shelf (or Kindle) of any prospective ArcGIS programmer who wants to jump in and start exploring the river. I will certainly use it myself and recommend it to my students and colleagues.

Jerry Davis

Geography Professor and Department Chair from San Francisco State University

About the Authors

Silas Toms is a certified GIS Professional and the author of the first edition of *ArcPy and ArcGIS*. President and founder of Loki Intelligent Corporation, a location information firm located in San Francisco, California, he is an expert in real-time geographic information systems and analysis automation. Along with Dara O'Beirne and Arini Geographics, he developed the real-time common operational picture used at Super Bowl 50 and all other events at Levi's Stadium in Santa Clara, California. This dynamic system was recognized by the White House and ESRI President, Jack Dangermond, as a unique and powerful application of GIS, allowing the federal, state, and local government to coordinate and communicate in real time, for the first time ever.

As the President of Loki Intelligent, Silas is focused on unique applications of GIS that will power the future of location information. The sheer amount of data collected through sensors and mobile reporting demands automation and data processing improvements to turn the raw input into location intelligence. He believes that correct application of geospatial analysis, web mapping, and mobile data collection will improve the decision-making processes within the government and business. Loki is location information, and information is power.

I would like to thank my girlfriend Maureen for her support. I would like to thank my father Bruce, mother Susan, step-father Bob, and my sister Ashley and her family for giving me great advice and support. I would like to thank my professors at Humboldt State University, San Franciso State University, and UA. Finally, I would like to thank my friends, both professional and personal, who helped me become a good person and great geographer.

Dara O'Beirne is a certified GIS Professional (GISP) with over 10 years of GIS and Python experience. He earned both his bachelors and masters of art in geography from the San Francisco State University. He is currently a GIS analyst working at the City of Sacramento's Department of Utilities. Before joining the City of Sacramento, he was a GIS analyst for Arini Geographics at the City of Santa Clara in California. While in Santa Clara, he worked on converting the utility network's AutoCAD data to ESRI's GIS Local Government Information Model. He was also an integral part of the GIS Team, which included Silas Toms, Cyrus Hiatt, Sherie Tan, and Gabriel Paun, who worked on developing a web mapping application used during each event at the new Levi's Stadium, including Super Bowl 50.

Dara's professional experience also includes working with Towill Inc., a GIS and land surveying company in Northern California. At Towill, he played a central role in developing and implementing procedures related to the collection and analysis of LiDAR data for environmental and engineering applications. Prior to Towill, Mr. O'Beirne gained his professional GIS experience working for the Golden Gate National Recreation Area managed by the National Park Service, one of the largest urban park systems in the world, which includes national treasures such as Alcatraz, Muir Woods, and the Marin Headlands.

I would like to thank my wife Kate and my daughters Anya and Bryn, and to tell them that I love them. I would like to thank my family who has always guided and supported me, from Ireland to America and beyond. I would like to thank my professors at San Francisco State University, and all of my colleagues and friends who have helped me along the way.

About the Reviewers

Ayodele Adeyemo is a geogeek working with eHealth Africa as a GIS specialist. He has over 4 years of experience working on various open source and proprietary software packages to deliver efficient data management and solutions across industries. He is passionate about leveraging on data and technology to proffer solutions to human, social, and environmental solutions.

Ayodele led the team that built the Nigeria Open Data Access and also won the GIS Cloud GIS Day Contest in 2015, where he worked with a team to build a malaria vulnerability predictive system based on environmental and climatic factors. He is currently working on an alternative addressing system to power and deliver smart cities technologies in communities around the world.

Feel free to contact him on `me@ayodeleadeyemo.com`.

Ken Doman is a Senior Frontend Engineer at GEO Jobe, a geographic information systems consultant company and ESRI business partner that helps public sector organizations and private sector businesses get the most out of geospatial solutions. He has worked for both municipal government and the private sector. He has experienced many facets of GIS technology, ranging from field data collection, mapping and data analysis, to creating and deploying web mapping applications and solutions.

Ken is the author of *Mastering ArcGIS Server Development with JavaScript*. He has also reviewed several books for Packt, including *Building Web and Mobile ArcGIS Server Applications with JavaScript* and *Spatial Analysis with ArcGIS* by Eric Pimpler, and *ArcGIS for Desktop Cookbook* by Daniela Christiana Docan.

I'd first like to thank my wife, Luann, who puts up with my late nights reviewing books like this. I'd like to thank my current employer, GEO Jobe GIS Consulting, as well as past employers like Bruce Harris and Associates, City of Plantation, FL and City of Jacksonville, TX for believing in me and letting me learn so much on the job. Finally, I'd like to thank my creator for putting me where I need to be.

www.PacktPub.com

For support files and downloads related to your book, please visit www.PacktPub.com.

Did you know that Packt offers eBook versions of every book published, with PDF and ePub files available? You can upgrade to the eBook version at www.PacktPub.com and as a print book customer, you are entitled to a discount on the eBook copy. Get in touch with us at service@packtpub.com for more details.

At www.PacktPub.com, you can also read a collection of free technical articles, sign up for a range of free newsletters and receive exclusive discounts and offers on Packt books and eBooks.

https://www.packtpub.com/mapt

Get the most in-demand software skills with Mapt. Mapt gives you full access to all Packt books and video courses, as well as industry-leading tools to help you plan your personal development and advance your career.

Why subscribe?

- Fully searchable across every book published by Packt
- Copy and paste, print, and bookmark content
- On demand and accessible via a web browser

Customer Feedback

Thanks for purchasing this Packt book. At Packt, quality is at the heart of our editorial process. To help us improve, please leave us an honest review on this book's Amazon page at `https://www.amazon.com/dp/1787282511`.

If you'd like to join our team of regular reviewers, you can e-mail us at `customerreviews@packtpub.com`. We award our regular reviewers with free eBooks and videos in exchange for their valuable feedback. Help us be relentless in improving our products!

Table of Contents

Preface   1

Chapter 1: Introduction to Python for ArcGIS   7

  Python as a programming language   8
    Interpreted language   8
    Standard (built-in) library   8
    Glue language   9
    Wrapper modules   9
  The basics of Python programming   9
    Import statements   10
    Variables   11
    For loops   12
    If/Elif/Else statements   12
    While statements   13
    Comments   13
  Data types   14
    Strings   14
    Integers   14
    Floats   15
    Data containers   15
    Zero-based indexing   15
    Lists   16
    Tuples   17
    Dictionaries   17
  Other important concepts   18
    Indentation   19
    Functions   19
    Keywords   19
    Namespaces   20
  Important Python modules   20
    The OS (operating system) module   20
    The sys (Python system) module   20
    The CSV, XLRD, and XLWT modules   21
    Commonly used built-in functions   21
    Standard library modules   21

How Python executes a script   22
  What is a Python script?   22
  Python interpreter   22
  Where is the Python interpreter located?   23
  Which Python interpreter should be used?   23
  How does the machine know where the interpreter is?   24
  Make Python scripts executable when clicked   26
Integrated Development Environments (IDEs)   27
  IDLE   27
  PythonWin   27
  Atom, Notepad++, and Sublime Text   28
  IDE summary   28
Python folder structure   28
  Where modules reside   28
  Installing a third-party module   29
  Using Python's sys module to add a module   30
  The sys.path.append method   30
Summary   31
Chapter 2: Creating the First Python Script   33
Prerequisites   33
ModelBuilder   34
  Creating a model and exporting to Python   34
    Modeling the Select and Buffer tools   35
    Adding in the Intersect tool   36
    Tallying the analysis results   37
Exporting the model and adjusting the script   38
  The Automatically generated script   38
  File paths in Python   40
String manipulation   41
  String manipulation method 1: string addition   41
  String manipulation method 2: string formatting #1   42
  String manipulation method 3: string formatting #2   43
The ArcPy tools   44
  The Intersect tool   45
    Adjusting the script   45
    Adding the CSV module to the script   45
  Accessing the data: using a cursor   46
    Exceptions and tracebacks   48
    Overwriting files   48
The final script   49

Summary	50
Chapter 3: ArcPy Cursors - Search, Insert, and Update	51
Python functions – avoid repeating code	52
Technical definition of functions	52
The first function	53
Functions with parameters	54
Using functions to replace repetitive code	55
The createCSV function	56
Creating an XLS using XLWT	56
The data access module	57
Search cursors	58
Attribute field interactions	60
Update cursors	61
Updating the shape field	62
Adjusting a point location	62
Deleting a row using an Update cursor	63
Using an Insert cursor	63
Inserting a polyline geometry	65
Inserting a polygon geometry	66
Summary	67
Chapter 4: ArcPy Geometry Objects and Cursors	69
ArcPy geometry object classes	70
ArcPy Point objects	71
ArcPy Array objects	72
ArcPy Polyline objects	73
ArcPy Polygon objects	75
Polygon object buffers	75
Other Polygon object methods	77
The AsShape method	78
Generic geometry object	79
ArcPy PointGeometry objects	79
Rewriting the bus stop analysis	80
Adding to the analysis	82
Summary	83
Chapter 5: Creating a Script Tool	85
Adding dynamic parameters to a script	85
Accessing the passed parameters	86
Displaying script messages using arcpy.AddMessage	88
Adding dynamic components to the script	88

Creating a script tool 90
Labeling and defining parameters 92
Adding data types 94
Adding the Bus Stop feature class 94
Adding the Census Block feature class 95
Adding the Census Block field 96
Adding the output spreadsheet 97
Adding the spreadsheet field names 97
Adding the SQL Statement 98
Adding the bus stop fields 98
Inspecting the final script 99
Running the script tool 101
Summary 102

Chapter 6: The arcpy.mapping Module 103
Using ArcPy with map documents 104
Interacting with map document elements 104
Data frames 104
Pan and zoom methods 105
Using the arcpy.mapping module to control layer objects 106
Layer object methods and properties 106
Data source 107
Name or description 107
Visibility 107
Definition queries 107
Inspecting and replacing layer sources 108
The ListBrokenDataSources method 108
Fixing the broken links 108
Fixing the links of individual layers 109
Exporting to PDF from an MXD 110
Automated map document production 110
The variables 113
Connection to the map document 113
Data frames 114
Access the layers 114
The layout elements 114
Generating a buffer from the bus stops feature class 115
Intersecting the bus stop buffer and census blocks 115
Format a dynamic definition query 116
Updating the layout elements 118
Exporting the adjusted map to PDF 118
The Results - Dynamic maps 119
Summary 120

Chapter 7: Advanced Analysis Topics 121

Using Network Analyst	121
Creating a network dataset	122
Importing the datasets	123
Creating the network dataset	124
Accessing the network dataset using ArcPy	125
Breaking down the script	125
The Network Analyst module	127
Accessing the Spatial Analyst extension	128
Adding elevation to the bus stops	128
Using Map algebra to generate elevation in feet	129
Adding in the bus stops and getting elevation values	130
The final result	130
Summary	131
Chapter 8: Introduction to ArcGIS Online	133
ArcGIS Online	133
Signing up for an account	134
Exploring the interface	136
The My Organization tab	137
The My Content tab	137
The Add Item option	138
Features from services	139
Features from files	139
The Create tab	140
The Groups tab	141
The Map and Scene tabs	142
Publishing from an MXD	142
Styling the layers	143
Publishing the layers	144
The Share As menu	146
Service Editor	148
The Item Description option	149
Analyze	150
Updates	151
Developer account	152
Summary	153
Chapter 9: ArcPy and ArcGIS Online	155
ArcGIS Online REST services	155
Exploring ArcGIS REST services	156
URL parameters	159
Feature sets	162
Feature set methods	163
ArcGIS Online tokens	168

Putting it all together	172
Summary	178
Chapter 10: ArcREST Python Package	**179**
Introducing the ArcREST module	180
Installing ArcREST	180
Introduction to the ArcREST package structure	181
ArcREST security handler	183
ArcGIS Online administration	184
Querying hosted feature services	185
Querying all features and saving as a feature class	185
Adding a field to a feature service	188
Adding domains to fields in a hosted feature service	191
Appending a feature class to a feature service	194
Updating records in a feature service	196
Summary	199
Chapter 11: ArcPy and ArcGIS Pro	**201**
Introducing ArcGIS Pro	202
Installing and configuring ArcGIS Pro	202
The ArcGIS Pro Python window	205
Python 2.7 and Python 3.5 with ArcPro	210
Conda and ArcGIS Pro	212
Running standalone scripts with Conda	213
Reviewing Conda basics	220
Summary	224
Chapter 12: ArcGIS API for Python	**225**
Introduction to the ArcGIS API for Python	225
Installing and configuring Anaconda with Jupyter	226
Install the ArcGIS Python API	230
Creating a Jupyter Notebook	231
Starting the ArcGIS API for Python	238
Adding an item to a web map	241
Importing a CSV with pandas	242
Summary	246
Index	**247**

Preface

The updated ArcPy and ArcGIS now cover three major Python modules for the automation of geospatial analysis and data administration. ArcPy, ArcREST, and the ArcGIS API for Python are all powerful Python libraries that allow GIS analysts and data scientists to script the processing and publication of location data. Using the power of programming with ArcGIS for Desktop, ArcGIS for Server, and ArcGIS Online has never been easier, or more important. With all new chapters and code, this book showcases how to use each module correctly to improve any geospatial workflow.

What this book covers

Chapter 1, *Introduction to Python for ArcGIS*, covers basic Python syntax and tools, and introduces the ArcPy code library along with other useful modules.

Chapter 2, *Creating the First Python Script*, takes the user from the ArcGIS ModelBuilder environment, where they model a geospatial analysis and export it as an ArcPy script.

Chapter 3, *ArcPy Cursors - Search, Insert, and Update*, explores the use of cursors or code tools used to programmatically create, update, and access location data.

Chapter 4, *ArcPy Geometry Objects and Cursors*, explains the use of cursors and geometry objects, which are ArcPy classes used to perform geospatial analysis in custom scripts.

Chapter 5, *Creating a Script Tool*, demonstrates how to create a custom ArcToolbox script tool with a graphical user interface that can be used like any other Esri tool.

Chapter 6, *The arcpy.mapping Module*, outlines the use of the arcpy.mapping module and its role in automated map production.

Chapter 7, *Advanced Analysis Topics*, discusses the use of ArcPy for advanced spatial and network analysis.

Chapter 8, *Introduction to ArcGIS Online*, takes you through signing up for ArcGIS Online, publishing data from an MXD, and creating an Esri Developers Account.

Chapter 9, *ArcPy and ArcGIS Online*, looks at the use of ArcPy with ArcGIS Online and the ArcGIS REST API for programmatic access to cloud-based location data.

Chapter 10, *ArcREST Python Package*, explores the use of the ArcREST Python package, which allows advanced control of the ArcGIS REST API.

Chapter 11, *ArcPy and ArcGIS Pro*, lists the new Python 3.5 libraries and the configuration of the software used to program analysis in ArcGIS Pro.

Chapter 12, *ArcGIS API for Python*, dives into the use of the new ArcGIS API for Python, which allows access to ArcGIS Online data within Jupyter notebooks.

Chapter 13, *Python and ArcGIS Enterprise,* teaches us how to use each of these three Python modules within a professional GIS workflow.

What you need for this book

This book uses Esri's ArcGIS software suite, including ArcGIS for Desktop, ArcGIS for Server, and ArcGIS Online, along with the Python programming language. Python and ArcPy are installed with the ArcGIS for Desktop software, which is available at `http://www.arcgis.com`. ArcREST is available for download from GitHub, and the ArcGIS API for Python is available at `https://developers.arcgis.com/python/`.

Who this book is for

This book is for anyone who works with location information. GIS analysts, data scientists, planners, biologists, web programmers, and students as well as many others can all benefit from the correct application of Python geospatial programming tools.

Conventions

In this book, you will find a number of text styles that distinguish between different kinds of information. Here are some examples of these styles and an explanation of their meaning.

Code words in text, database table names, folder names, filenames, file extensions, pathnames, dummy URLs, user input, and Twitter handles are shown as follows:

"The `createCSV` function accepts the dictionary and the name of the output CSV file as strings."

A block of code is set as follows:

```
def firstFunction():
 'a simple function returning a string'
 stringvariable = "My First Function"
 return stringvariable

>>> firstFunction()
```

When we wish to draw your attention to a particular part of a code block, the relevant lines or items are set in bold:

```
def firstFunction():
 'a simple function returning a string'
 stringvariable = "My First Function"
 return stringvariable

>>> firstFunction()
```

Any command-line input or output is written as follows:

```
conda install -c esri arcgis
```

New terms and **important words** are shown in bold. Words that you see on the screen, for example, in menus or dialog boxes, appear in the text like this: "Once complete, you should see the hosted feature layer under **My Content**, as shown here."

Warnings or important notes appear in a box like this.

Tips and tricks appear like this.

Reader feedback

Feedback from our readers is always welcome. Let us know what you think about this book-what you liked or disliked. Reader feedback is important for us as it helps us develop titles that you will really get the most out of. To send us general feedback, simply e-mail feedback@packtpub.com, and mention the book's title in the subject of your message. If there is a topic that you have expertise in and you are interested in either writing or contributing to a book, see our author guide at www.packtpub.com/authors.

Customer support

Now that you are the proud owner of a Packt book, we have a number of things to help you to get the most from your purchase.

Downloading the example code

You can download the example code files for this book from your account at http://www.p acktpub.com. If you purchased this book elsewhere, you can visit http://www.packtpub.c om/supportand register to have the files e-mailed directly to you. You can download the code files by following these steps:

1. Log in or register to our website using your e-mail address and password.
2. Hover the mouse pointer on the **SUPPORT** tab at the top.
3. Click on **Code Downloads & Errata**.
4. Enter the name of the book in the **Search** box.
5. Select the book for which you're looking to download the code files.
6. Choose from the drop-down menu where you purchased this book from.
7. Click on **Code Download**.

Once the file is downloaded, please make sure that you unzip or extract the folder using the latest version of:

- WinRAR / 7-Zip for Windows
- Zipeg / iZip / UnRarX for Mac
- 7-Zip / PeaZip for Linux

The code bundle for the book is also hosted on GitHub at https://github.com/PacktPubl ishing/ArcPy-and-ArcGIS-Second-Edition. We also have other code bundles from our rich catalog of books and videos available at https://github.com/PacktPublishing/. Check them out!

Downloading the color images of this book

We also provide you with a PDF file that has color images of the screenshots/diagrams used in this book. The color images will help you better understand the changes in the output. You can download this file from https://www.packtpub.com/sites/default/files/downloads/ArcPyandArcGISSecondEdit ion_ColorImages.pdf.

Errata

Although we have taken every care to ensure the accuracy of our content, mistakes do happen. If you find a mistake in one of our books-maybe a mistake in the text or the code-we would be grateful if you could report this to us. By doing so, you can save other readers from frustration and help us improve subsequent versions of this book. If you find any errata, please report them by visiting http://www.packtpub.com/submit-errata, selecting your book, clicking on the **Errata Submission Form** link, and entering the details of your errata. Once your errata are verified, your submission will be accepted and the errata will be uploaded to our website or added to any list of existing errata under the Errata section of that title. To view the previously submitted errata, go to https://www.packtpub.com/books/content/support and enter the name of the book in the search field. The required information will appear under the **Errata** section.

Piracy

Piracy of copyrighted material on the Internet is an ongoing problem across all media. At Packt, we take the protection of our copyright and licenses very seriously. If you come across any illegal copies of our works in any form on the Internet, please provide us with the location address or website name immediately so that we can pursue a remedy. Please contact us at copyright@packtpub.com with a link to the suspected pirated material. We appreciate your help in protecting our authors and our ability to bring you valuable content.

Questions

If you have a problem with any aspect of this book, you can contact us at questions@packtpub.com, and we will do our best to address the problem.

1
Introduction to Python for ArcGIS

In this chapter, we will discuss the development of Python as a programming language from its introduction in the late 1980s to its current state. We will discuss the creator of the language and the philosophy of design that spurred its development. We will also touch on important modules that will be used throughout the rest of the book, and especially focus on the modules built into the Python Standard Library. We will configure Python and the computer to execute Python scripts. The structure of the Python folder will be discussed, as will the location of the ArcPy module within the ArcGIS folder structure. We will also discuss **Integrated Development Environments (IDEs)**--programs designed to assist in code creation and code execution--comparing and contrasting existing IDEs to determine what benefits each IDE can offer when scripting Python code.

This chapter will cover the following topics:

- Quick Overview of Python: what it is and what it does, who created it, where is it now
- Important Python modules, both built-in and third party
- Python core concepts including data types, data containers, and looping
- The location of the Python interpreter, and how it is called to execute a script
- Adjusting the computer's environment variables to ensure correct code execution
- Integrated Development Environments (IDEs)
- Python's folder structure, and where the modules are stored

Python as a programming language

Over the last 40+ years of computer science, programming languages have developed from assembly and machine code towards high-level abstracted languages, which are much closer to English. The Python programming language, begun by Guido van Rossum in 1989, was designed to overcome issues that programmers were dealing with in the 1980s: slow development time, overly complicated syntax, and horrible readability. He developed Python as a language that would enable rapid code development and testing, have beautiful (or at least readable) syntax, and produce results with fewer lines of code, in less time. The first version of Python (0.9.0) was released in 1991, and it has always been free to download and use.

 Go to `https://www.python.org/` to explore Python documentation, try tutorials, get help, find useful Python code libraries, and download Python. Python has multiple major and minor versions. For much of the book, we are using Python 2.7, which is installed automatically along with ArcGIS for Desktop. For chapters on ArcGIS Pro, we will use Python 3.5.

Interpreted language

Python is an interpreted language. It is written in C, a compiled language, and the code is "interpreted" from Python into C before it is executed. Practically, this means that the code is executed as soon as it is converted and compiled. While the code interpretation can have speed implications for the execution of Python-based programs, this has very little real-world implications for its use with ArcGIS. Testing of code snippets is much faster in an interpretive environment, and it is perfect for creating scripts to automate basic, repeatable computing tasks.

Standard (built-in) library

Python, when installed, has a basic set of functions that are referred to as the built-in library. These built-in tools allow Python to perform string manipulations, math computations, HTTP calls, and URL parsing along with many other functions. Some of the tool libraries, or modules, are available as soon as Python is started, while others must be explicitly called using the "import" keyword to make their functions and classes available. Other modules have been developed by third parties, and can be downloaded and installed onto the Python installation as needed.

Glue language

Python is often called a "glue" language. This term describes the use of Python code to control other software programs by sending inputs to their **Application Programming Interface** (**API**) and collecting outputs, which are then sent to another program to repeat the process. A GIS example would be to use Python's `urllib2` to download zipped shapefiles from a website, unzipping the files, processing the files using ArcToolbox, and compiling the results into an Excel spreadsheet. All of this is accomplished using freely available modules that are either included in Python's built-in library, or added on when ArcGIS is installed.

Wrapper modules

The ArcPy module is a "wrapper" module. It "wraps" a Python code interface over the existing ArcGIS tools and source code, allowing us to access these tools within our scripts. Wrapper modules are common in Python, and are so named because they "wrap" Python onto the tools we will need. They allow us to use Python to interface with programs written in C or other programming languages using the API of those programs. The wrapper module `os` allows for operating system operations.

For example, wrapper modules make it possible to extract data from an Excel spreadsheet, transform the data into another format such as a shapefile, and load it into an MXD as a layer. Not all modules are wrappers; some modules are written in "pure Python", and perform their analysis and computations using Python syntax. Either way, the end result is that a computer and its programs are available to be manipulated and controlled using Python.

The basics of Python programming

Python has a number of language requirements and conventions, which allow for the control of modules and the structuring of code. Following are a number of important basic concepts which will be used throughout this book, and when crafting scripts for use with geospatial analysis.

To test these examples, open the `IDLE (Python GUI)` program from the `Start Menu/ArcGIS/Python2.7` folder after installing ArcGIS for Desktop. It has a built-in "interpreter" or code entry interface, indicated by the triple chevron >>> and a blinking cursor. To create a script in IDLE to save your code, click on to the **File** menu and then click **New File**. Save any script with a `.py` extension. Otherwise, just enter commands into the interpreter and push *Enter* to execute or add the next line.

Import statements

Import statements are used to augment the power of Python by calling other modules for use in the script. These modules can be a part of the standard Python library of modules, such as the `math` module (used to do higher mathematical calculations), or, importantly, can be like ArcPy, which will allow us to interact with ArcGIS. Import statements can be located anywhere before the module is used, but, by convention, they are located at the top of a script.

There are three ways to create an import statement. The first, and most standard, is to import the whole module as follows:

```
import arcpy
```

Using this method, we can even import more than one module on the same line. Next, we will import three modules: `arcpy, os` (the operating system module), and `sys` (the Python system module):

```
import arcpy, os, sys
```

The next method of importing a script is to import a specific portion of a module instead of importing the entire module using the `from <module> import <submodule>` syntax:

```
from arcpy import mapping
```

This method is used when only a portion of the code from ArcPy is needed; it has the practical effect of limiting the amount of memory used by the module when it is called. We can also import multiple portions of the module in the same fashion.

```
from arcpy import mapping, da
```

The third way to import a module is to write the `from <module> import <submodule>` syntax, but use an asterisk * to import all parts of the module as follows:

```
from arcpy import *
```

This method is still used, but it is discouraged as it can have unknown effects--the main one is that the names of the variables in the module might conflict with another variable in another module. For this reason, it is best to avoid this third method. However, lots of existing scripts include import statements in this format, so it is good to know that it exists.

Variables

Variables are a part of all programming languages. They are used to reference data objects stored in memory for using later in a script. There are a lot of arguments over the best method of naming variables. No variable standard has been developed for Python scripting for ArcGIS, so I will describe some common practices to use when naming variables here:

1. Making them descriptive: Don't just name a variable, *x*; that variable will be useless later when the script is reviewed, and there is no way of knowing what it is used for, or why. They should be longer rather than shorter, and should explain what they do, or even what type of data they hold. For example:

   ```
   shapefilePath = "C:/Data/shapefile.shp"
   ```

2. Using `CamelCase` to make the variable readable: Camel case is a term used for variables that start with a lowercase letter but have uppercase letters in the middle, resembling a camel's hump. For example:

   ```
   camelCase = 'camel case is twoWords stuck together like this'
   ```

3. Using an underscore to separate parts of the name: This makes the name longer, but adds some clarity when reading the variable name, like this:

   ```
   location_address = '100 Main St'
   ```

4. Including the data type in the variable name: If the variable contains a string, call it `variableString` or `variable_string`. This is not standard, and will not be used in this book, but it can help organize the script, and is helpful for others who will read these scripts. Python is dynamically typed instead of statically typed, a programming language distinction, which means that a variable does not have to be declared before it can be used, unlike Visual Basic or other statically typed languages. For example:

   ```
   variableString = 'this is a string'
   ```

For loops

Built into all programming languages is the ability to iterate over a dataset to perform an operation on the data, thus transforming the data or extracting data that meets specific criteria. The dataset must be iterable to be used in a `for` loop. We will use iteration in the form of `for` loops throughout this book. Here is a simple example of a `for` loop, which takes string values and prints them in uppercase using the string `upper` method. Open IDLE (Python GUI) from the `Start Menu/ArcGIS/Python2.7` folder to try a `for` loop. Enter commands at the Python interpreter's triple chevron >>> :

```
>>> newlist = ['a','b','c','d']
>>>for value in newlist:
        print value.upper()
A
B
C
D
```

If/Elif/Else statements

Conditional statements, called `if...else` statements in Python, are another programming language standard. They are used when evaluating data; when certain conditions are met, one action will be taken (the initial `if` statement); if another condition is met, another action is taken (this is an `elif` statement), and if the data does not meet the condition, a final action is assigned to deal with those cases (the `else` statement). They are similar to a conditional in an SQL statement used with the Select Tool in **ArcToolbox**. Here is an example using an `if...else` statement to evaluate data in a list. In the example, within the `for` loop, the modulo operator % produces the remainder of a division operation. The `if` condition checks for no remainder when divided in half, a `elif` condition looks for remainder of two when divided by three, and the `else` condition catches any other result, as shown:

```
>>> data = [1,2,3,4,5,6,7]
>>> for val in data:
        if val % 2 == 0:
            print val,"no remainder"
        elif val % 3 == 2:
            print val, "remainder of two"
        else:
            print "final case"


final case
```

```
2 no remainder
4 no remainder
5 remainder of two
6 no remainder
final case
```

While statements

Another important evaluation tool is the `while` statement. It is used to perform an action while a condition is true; when the condition is false, the evaluation will stop. Note that the condition must become false, or the action will be performed forever, creating an "infinite loop" that will not stop until the Python interpreter is shut off externally. Here is an example of using a `while` loop to perform an action until a true condition becomes false:

```
>>> x = 0
>>> while x < 5:
        print x
        x+=1
0
1
2
3
4
```

Comments

Comments in Python are used to add notes within a script. They are marked by a pound sign, and are ignored by the Python interpreter when the script is run. Comments are useful for explaining what a code block does when it is executed, or for any other helpful note that a script author would like to make for future script users:

```
#This is a comment
```

Data types

GIS uses points, lines, polygons, coverages, and rasters to store data. Each of these GIS data types can be used in different ways when performing analyses, and have different attributes and traits. Python, like GIS, has data types that it uses to organize data. The main data types used in this book are strings, integers, floats, lists, tuples, and dictionaries. They each have their own attributes and traits, and are used for specific parts of code automation. There are also built-in functions that allow for data types to be converted from one type to another; for instance, the integer 1 can be converted to the string '1' using the function str():

```
>>> variable = 1
>>> strvar = str(variable)
>>> strvar
'1'
```

Strings

Strings are used to contain any kind of character, and begin and end with quotation marks. Either single or double quotes can be used; the string must begin and end with the same type of quotation mark. Quoted text can appear within a string; it must use the opposite quotation mark to avoid conflicting with the string, as shown here:

```
>>> quote = 'This string contains a quote: "Here is the quote" '
```

A third type of string is also employed: a multiple line string, which starts and ends with three single quote marks like this:

```
>>> multiString = '''This string has
multiple lines and can go for
as long as I want it too'''
```

Integers

Integers are whole numbers that do not have any decimal places. There is a special consequence to the use of integers in mathematical operations: if integers are used for division, an integer result will be returned. Check out the following code snippet to see an example of this:

```
>>> 5/2
2
```

Instead of an accurate result of 2.5, Python will return the "floor", or the lowest whole integer for any integer division calculation. This can obviously be problematic, and can cause small bugs in scripts, which can have major consequences. Please be aware of this issue when writing scripts.

Floats

Floating point values, or floats, are used by Python to represent decimal values. Because computers store values in a base 2 binary system, there can be issues representing a floating value that would normally be represented in a base 10. Read `https://docs.python.org /2/tutorial/floatingpoint.html`for a further discussion on the ramifications of this limitation; for applications discussed within this book, it won't be an issue.

Data containers

Data must often be grouped, ordered, counted, and sorted. Python has a number of built-in data "containers", which can be used for each and all of these needs. Lists, tuples, sets, and dictionaries are the main data containers, and can be created and manipulated without the need to import any libraries.

For array types like lists and tuples, the order of the data is very important for retrieval. Data containers like dictionaries "map" data using a "key-value" retrieval system, where the "key" is mapped or connected to the "value". In dictionaries, the order of the data is not important. For all mutable data containers, sets can be used to retrieve all unique values within a data container such as a list.

Zero-based indexing

Data stored in ordered arrays like lists and tuples often needs to be individually accessed. To directly access a particular item within a list or tuple, you need to pass its index number to the array in square brackets. This makes it important to remember that Python indexing and counting starts at 0 instead of 1. This means that the first member of a group of data is at the 0 index position, and the second member is at the 1 index position, and so on:

```
>>> newlist = ['run','chase','dog','bark']
>>> newlist[0]
'run'
>>> newlist[2]
'dog'
```

Zero-based indexing applies to characters within a string. Here, the list item is accessed using indexing, and then individual characters within the string are accessed, also using indexing:

```
>>> newlist[3][0]
'b'
```

Zero-based indexing also applies when there is a `for` loop iteration within a script. When the iteration starts, the first member of the data container being iterated is data item 0, the next is data item 1, and so on:

```
>>> newlist = ['run','chase','dog','bark']
>>> for counter, item in enumerate(newlist):
        print counter, newlist[counter]


0 run
1 chase
2 dog
3 bark
```

The `enumerate` module is used to add a counter variable to a `for` loop, which can report the current index value.

Lists

Lists are ordered arrays of data, which are contained in square brackets, `[]`. Lists can contain any other type of data, including other lists. Mixing of data types, such as floats, integers, strings, or even other lists, is allowed within the same list. Lists have properties, such as length and order, which can be accessed to count and retrieve. Lists have methods to be extended, reversed, sorted, and can be passed to built-in Python tools to be summed, or to get the maximum or minimum value of the list.

Data pieces within a list are separated by commas. List members are referenced by their index or position in the list, and the index always starts at zero. Indexes are passed to square brackets `[]` to access these members, as in the following example:

```
>>> alist = ['a','b','c','d']
>>> alist[0]
'a'
```

This preceding example shows us how to extract the first value (at index 0) from the list called `alist`. Once a list has been populated, the data within it is referenced by its index, which is passed to the list in square brackets. To get the second value in a list (the value at index 1), the same method is used:

```
>>> alist[1]
'b'
```

Lists, being mutable, can be changed. Data can be added or removed. To merge two lists, the `extend` method is used:

```
>>> blist = [2,5,6]
>>> alist.extend(blist)
>>> alist
['a', 'b', 'c', 'd', 2, 5, 6]
```

Lists are a great way to organize data, and are used all the time in ArcPy.

Tuples

Tuples are cousins to lists, and are denoted by parentheses `()`. Unlike lists, tuples are "immutable". No data can be added or removed, nor can they cannot be adjusted or extended, once a tuple has been created in memory (it can be overwritten). Data within a tuple is referenced in the same way as a list, using index references starting at 0 passed to square brackets `[]`:

```
>>> atuple = ('e','d','k')
>>> atuple[0]
'e'
```

Dictionaries

Dictionaries are denoted by curly brackets "{ }", and are used to create "key-value" pairs. This allows us to map values from a key to a value so that the value can replace the key, and data from the value can be used in processing. Here is a simple example:

```
>>> new_dic = {}
>>> new_dic['key'] = 'value'
>>> new_dic
{'key': 'value'}
>>> adic = {'key':'value'}
>>> adic['key']
'value'
```

Note that instead of referring to an index position, like lists or tuples, the values are referenced using a key. Also, keys can be any data object except lists (because they are mutable).

This can be very valuable when reading data from a shapefile or feature class for use in analysis. For example, when using an **address_field** as a key, the value would be a list of row attributes associated with that address. Look at the following example:

```
>>> business_info = { 'address_field' :   ['100', 'Main', 'St'],
'phone':'1800MIXALOT'   }
>>> business_info['address_field']
['100', 'Main', 'St']
```

Dictionaries are very valuable for reading in feature classes, and for easily parsing through the data by calling only the rows of interest, among other operations. They are great for ordering and reordering data for later use in a script.

Dictionaries are also useful for counting data items such as the number of times a value appears within a dataset, as seen in this example:

```
>>> list_values = [1,4,6,7,'foo',3,2,7,4,2,'foo']
>>> count_dictionary = {}
>>> for value in list_values:
        if value not in count_dictionary:
            count_dictionary[value] = 1
        else:
            count_dictionary[value] += 1
>>> count_dictionary['foo']
2
```

Other important concepts

The use of Python for programming requires an introduction to a number of concepts that are either unique to Python, but required, or common programming concepts that will be invoked repeatedly when creating scripts. The following are a number of these concepts which must be covered to be fluent in Python.

Indentation

Python, unlike most other programming languages, enforces strict rules on indenting lines of code. This concept derives again from Guido's desire to produce clean, readable code. When creating functions, or using `for` loops or `if...else` statements, indentation is required on the succeeding lines of code. If a `for` loop is included inside an `if...else` statement, there will be two levels of indentation. New programmers generally find it to be helpful, as it makes it easy to organize code. A lot of programmers new to Python will create an indentation error at some point, so make sure to pay attention to the indentation levels.

Functions

Functions are used to take code that is repeated over and over within a script, or across scripts, and make formal tools out of them. Using the keyword `def`, short for "define function", functions are created with defined inputs and outputs (which are returned from the function using the keyword `return`). The idea of a function in computing is that it takes in data in one state, and converts it into data in another state, without affecting any other part of the script. This can be very valuable for automating a GIS analysis.

Here is an example of a function that returns the square of any number supplied:

```
def square(inVal):
    return inVal ** 2
>>> square(3)
9
```

Keywords

There are a number of keywords built into Python that should be avoided when naming variables. These include `max`, `min`, `sum`, `return`, `list`, `tuple`, `def`, `del`, `from`, `not`, `in`, `as`, `if`, `else`, `elif`, `or`, and `while` among many others. Using these keywords as variables can result in errors in your code.

Namespaces

Namespaces are a logical way to organize variable names, to allow a variable inside a function or an imported module to share the same name as a variable in the main script body, without overwriting the variable. Referred to as "local" variables versus "global" variables, local variables are contained within a function (either in the script or within an imported module), while global variables are within the main body of the script.

These issues often arise when a variable within an imported module unexpectedly has the same name of a variable in the script, and the interpreter has to use namespace rules to decide between the two variables.

Important Python modules

Modules, or code libraries that can be called by a script to increase its programming potential, are either built into Python, or are created by third parties, and added later to Python. Most of these are written in Python, but a number of them are also written in other programming languages, and then "wrapped" in Python to make them available within Python scripts. Wrappers are also used to make other software available to Python, such as the tools built into Microsoft Excel.

The OS (operating system) module

The os module, part of the standard library, is very useful for a number of regular operations within Python. The most used part of the os module is the os.path method, which allows the script to control file paths, and to divide them into directory paths and base paths. There is also a useful method, os.walk, which will "walk" a directory and return all files within the folders and the subfolders.

The sys (Python system) module

The sys module, part of the standard library, refers to the Python installation itself. It has a number of methods that will get information about the version of Python installed, as well as information about the script, and any "arguments" supplied to the script, using the sys.argv method. The sys.path method is very useful for appending the Python file path; practically, this means that folders containing scripts can be referenced by other scripts to make the functions they contain importable.

The CSV, XLRD, and XLWT modules

The csv, xlrd, and xlwt modules are used to read and write data spreadsheets. They can be very useful for extracting data from the spreadsheets and converting them into data for GIS analysis, or for writing out analysis results as spreadsheets when an analysis is complete. The csv module (which creates text file spreadsheets using text delimiters like commas) is a built-in module, while xlrd and xlwt (which read and write Excel files respectively) are not part of the standard library, but are installed along with ArcGIS and Python 2.7.

Commonly used built-in functions

There are a number of built-in functions that we will use throughout the book. The main ones are listed as follows:

- str: The string function is used to convert any other type of data into a string.
- int: The integer function is used to convert a string or float into an integer. To avoid an error, any string passed to the integer function must be a number such as '1'.
- float: The float function is used to convert a string or an integer into a float, much like the integer function.

Standard library modules

Commonly used standard library modules that must be imported are as follows:

- datetime: The datetime module has date and time information, and can convert date data formats
- math: The math module is for higher level math functions, such as getting a value for *Pi* or squaring a number
- string: The string module is used for string manipulations
- csv: The csv module is used for creating, accessing, and editing text spreadsheets.

Check out https://docs.python.org/2/library/ for a complete list of the built-in modules.

How Python executes a script

Understanding how Python works to interpret a script, and then executes the commands within, is as important as understanding the Python language itself. Hours of debugging and error checking can be avoided by taking the time to set up Python correctly. The interpretive nature of Python means that a script will first have to be converted into bytecode before it can be executed. We will cover the steps that Python takes to achieve our goal of automating GIS analysis.

What is a Python script?

Let's start with the very basics of writing and executing a Python script. What is a Python script? It is a simple text file that contains a series of organized commands, written in a formalized language. The text file has the extension .py, but other than that, there is nothing to distinguish it from any other text file. It can be opened using a text editor such as Notepad or WordPad, but the "magic" that Python does is that it does not reside in a Python script. Without the Python interpreter, a Python script cannot run, and its commands cannot be executed.

Python interpreter

The Python interpreter, in a Windows environment, is a program that has been 'compiled' from the Python source code into a Windows executable and has the extension .exe. The Python interpreter, python.exe, is written in C, an older and extensively used programming language with a more complex syntax.

The Python interpreter, as its name implies, interprets the commands contained within a Python script. When a Python script is run, or executed, the syntax is first checked to make sure that it conforms to the rules of Python (for example, indentation rules are followed, and that the variables follow naming conventions). Then, if the script is valid, the commands contained within are converted into bytecode, a specialized code that is executed by the bytecode interpreter, a virtual machine written in C. The bytecode interpreter further converts the bytecode (which is contained within files that end with the extension .pyc) into the correct machine code for the computer being used, and then the CPU executes the script. This is a complex process, which allows Python to maintain a semblance of simplicity.

Where is the Python interpreter located?

The location of the Python interpreter within the folder structure of a computer is an important detail to master. Python is often downloaded directly from `https://www.python.org/`, and installed separately from ArcGIS. However, each ArcGIS version will require a specific version of Python; given this requirement, the inclusion of Python within the ArcGIS installation package is helpful. For this book, we will be using ArcGIS 10.5, and this will require Python 2.7.

On a Windows machine, the Python folder structure is placed directly in the `C:\` drive unless it is explicitly loaded on another drive. The installation process for ArcGIS 10.5 will create a folder at `C:\Python27`, which will contain another folder called either `ArcGIS10.5` or `ArcGIS10.5x64` depending on the version of ArcGIS that has been installed. For this book, we will be using the 32-bit version of ArcGIS, so, the final folder path will be `C:\Python27\ArcGIS10.5`.

Within that folder are a number of subfolders as well as `python.exe`, which is the Python interpreter itself. Also included is a second version of the interpreter called `pythonw.exe`; this version is also very important, as it will execute a script without causing a terminal window to appear. Both `python.exe` and `pythonw.exe` contain complete copies of all Python commands, and can be used to execute a script.

Which Python interpreter should be used?

The general rule for executing a script directly using the Python interpreters is to use `pythonw.exe`, as no terminal window will appear. When there is a need to test code snippets, or to see output within a terminal window, then start `python.exe` by double-clicking the executable file.

When `python.exe` is started, a Python interpreter console will appear as seen in the following screenshot:

```
C:\Python27\python.exe

Python 2.7.12 (v2.7.12:d33e0cf91556, Jun 27 2016, 15:24:40) [MSC v.1500 64 bit (
AMD64)] on win32
Type "help", "copyright", "credits" or "license" for more information.
>>> _
```

Note the distinctive three chevrons >>> that appear below the header explaining version information. That is the Python "prompt" where code is entered to be executed line by line, instead of in a completed script. This direct access to the interpreter is useful for testing code snippets and understanding syntax. A version of this interpreter, the Python Window, has been built into ArcMap and ArcCatalog since ArcGIS 10. It will be discussed further in later chapters.

How does the machine know where the interpreter is?

To be able to execute Python scripts directly (that is, to make the scripts run by double-clicking on them), the computer will also need to know where the interpreter sits within its folder structure. To accomplish this requires both administrative account access, and advanced knowledge of how Windows searches for a program. If you have this, you can adjust an environment variable within the advanced system settings dialogue to register the interpreter with the system path.

On a Windows 7/10 machine, click on the Start menu, and right-click on **Computer.** Then select Properties from the menu. On a Windows 8 machine, open up Windows explorer, right click on **This PC**, and select **Properties** from the menu. These commands are shortcuts to get to the Control Panel's System and Security/System menu. Select Advanced system settings from the panel on the left. Click on the **Environment Variables** button at the bottom of the System Properties menu that appears. In the lower portion of the Environment Variables menu, scroll in the System variables window until the Path variable appears. Select it by clicking on it, and click on the **Edit** button. The Edit System Variable window will appear like this:

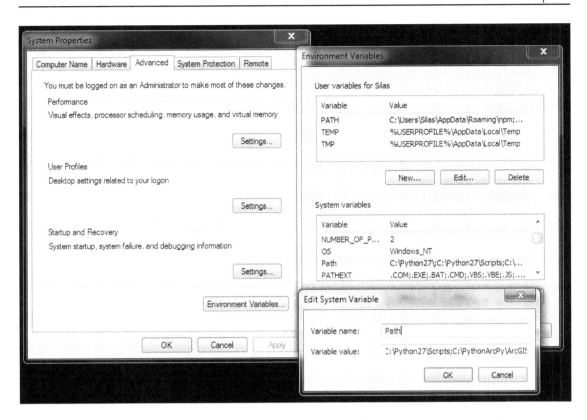

This variable has two components: **Variable name** (**Path**) and **Variable value**. The value is a series of folder paths separated by semicolons. This is the path that is searched when Windows looks for specific executables that have been associated with a file extension. In our case, we will add in the folder path that contains the Python interpreter. Type `C:\Python27\ArcGIS10.5` (or the equivalent on your machine) into the Variable value field, making sure to separate it from the value before it with a semi-colon. Press **OK** to exit the **Edit** dialogue, **OK** to exit the Environment Variables menu, and **OK** to exit the System Properties menu. The machine will now know where the Python interpreter is, as it will search all folders contained within the Path variable to look for an executable called Python. To test that the path adjustment worked correctly, open up a command window (`Start Menu/Run`, and type "cmd"), and type `python`.

The interpreter should start directly in the command window:

If the Python header with version information and the triple chevron appears, the path adjustment has worked correctly.

If there is no admin access available, there is a workaround. In a command-line window, pass the entire path to the Python interpreter (for example, `C:\Python27\ArcGIS10.5\python.exe`) to start the interpreter.

Make Python scripts executable when clicked

The final step to make the scripts run when clicked (which also means they can run outside of the ArcGIS environment, saving lots of memory overhead) is to associate files with the `.py` extension with the Python interpreter. If the scripts have not already been associated with the interpreter, they will appear as files of an unknown type or as a text file.

To change this, right-click on a Python script. Select **Open With**, and then select **Choose Default Program**. If `python.exe` or `pythonw.exe` does not appear as a choice, navigate to the folder that holds them (`C:\Python27\ArcGIS10.5` in this case), and select either python.exe or pythonw.exe. Again, the difference between the two is the appearance of a terminal window when the scripts are run using `python.exe`, which will contain any output from the script (but this window will disappear when the script is done). I recommend using `pythonw.exe` when executing scripts, and `python.exe` for testing out code. Python scripts can also explicitly call `pythonw.exe` by saving the script with the extension `.pyw` instead of `.py`.

Integrated Development Environments (IDEs)

The Python interpreter contains everything that is needed to execute a Python script or to test Python code by interacting with the Python interpreter. However, writing scripts requires a text editor. There are usually at least two included simple text editors on a Windows machine (Notepad and WordPad), and they would work in an emergency to edit a script or even write a whole script.

Unfortunately, they are very simple, and do not allow the user functionality that would make it easier to write multiple scripts or very long scripts. To bridge the gap, a series of programs, collectively known as Integrated Development Environments (IDEs), have been developed. IDEs exist for all programming languages, and include functions such as variable listing, code assist, and more, which makes them ideal for crafting programming scripts. We will review a few of them later to assess their usefulness for writing Python scripts. The following three discussed are all free and well-established within different Python communities.

IDLE

Python includes an IDE when it is installed. To start it in Windows 7, go to the `Start` menu, and find the `ArcGIS` folder within the `Programs` menu. Then find the `Python` folder; `IDLE` will be one of the choices within that folder. Select it to start IDLE.

IDLE contains an interactive interpreter (that is, the famous triple chevron), and the ability to run whole Python scripts. It is also written using Python's built-in GUI module called *Tkinter*, so it has the advantage of being written using the same language that it executes.

IDLE is a passable IDE, which is useful if no other programs can be installed on the machine. It is also very useful for rapid testing of code snippets. While it is not my IDE of choice, I find myself using IDLE almost daily.

PythonWin

PythonWin includes an Interactive Window where the user can directly interact with the Python interpreter. Scripts can also be opened within PythonWin, and it includes a set of tiling commands in the Windows menu, which allows the user to organize the display of all open scripts and the Interactive Window. It is very popular for users of ArcPy and ArcGIS, but it has been eclipsed in use by the full-fledged IDEs described as follows.

Atom, Notepad++, and Sublime Text

Some text editors have full-fledged code editing capabilities, making them ideal IDEs. While Sublime Text is commercial, it is a powerful program, which allows for easy code editing across multiple operating systems. Similarly, Notepad++ for Windows is a powerful text editor that works well for editing code. Atom, available at `https://atom.io/`, is a product of the GitHub development team, and offers multiple powerful language options such as code completion and error highlighting.

All three of these advanced text editors recognize Python keywords and code structure, and will make it easy to indent code according to the rules of Python. I use them all, often interchangeably, and have no strong opinion about which one is better, though I prefer Atom and Sublime Text, as these can be used in multiple operating systems, while Notepad++ is only available for Windows. They are powerful IDEs, which are available for download from online sources.

IDE summary

There are many other IDEs, both commercial and free, available for coding in Python. In the end, each GIS analyst must choose the tool that makes them feel productive and comfortable. This may change as programming becomes a bigger part of their daily work flow. Be sure to test out a few different IDEs to find one that is easy to use and intuitive.

Python folder structure

Python's folder structure holds more than just the Python interpreter. Within the subfolders reside a number of important scripts, digital link libraries, and even C language modules. Not all of the scripts are used all the time, but each has a role in making the Python programming environment possible. The most important folder to know about is the `site-packages` folder, where most modules that will be imported in Python scripts are contained.

Where modules reside

Within every Python installation is a folder called `Lib`, and within that folder is a folder called `site-packages`. On my machine, the folder sits at `C:\Python27\ArcGIS10.5\Lib\site-packages`.

Almost all third-party modules are copied into this folder to be imported as needed. The main exception to this rule, for our purposes, is the ArcPy module, which is stored within the `ArcGIS` folder in the `Program Files` folder (for example, `C:\Program Files (x86)\ArcGIS\Desktop10.5\arcpy`). To make that possible, the ArcGIS installer adjusts the Python system path (using the `sys` module) to make the `arcpy` module importable, as described next.

Installing a third-party module

To add greater functionality, thousands of third-party modules, or packages, are available for download. Online module repositories include the **Python Package Index** (**PyPI**) as well as GitHub, and others. Python 2 and Python 3 now include a module designed to make installing these packages more simple than it was in the past. This module, `pip`, will check for registered modules in PyPI, and install the latest version using the command `install`. Use `pip` from the command prompt by passing the command `install` and the name of the package to install.

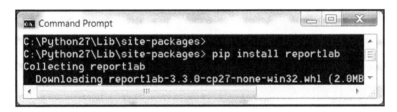

If the module is not available on PyPI, `pip` may not be able to install it. For instance, if it's on GitHub instead (even Python 3.7 is now hosted on `https://github.com/`, so GitHub is worth knowing), download the zip file of the module, and unzip it into the `Lib/site-packages` folder. Open a Command Prompt terminal, change directory (`cd`) into the newly unzipped folder, and run the script `setup.py` that is part of each module, using the command `python setup.py install`. This script will install the module, and configure the environmental variables required to make it run.

Many Python modules are only available in the GZip format, which can be unzipped using freeware such as 7Zip. Unzip the `.gz` file, then unzip the resulting `.tar` file into the `Lib/site-packages` folder in the Python folder.

Using Python's sys module to add a module

Python's `sys` module allows the user to take advantage of the system tools built into the Python interpreter. One of the most useful properties of the `sys` module is `sys.path`. It is a list of file paths which the user can modify to adjust where Python will look for a module to import, without needing administrative access.

When Python 2.7 is installed by the ArcGIS 10.5 installer, the installer takes advantage of `sys.path` functions to add `C:\Program Files (x86)\ArcGIS\Desktop10.5\arcpy` to the system path. To test this, start up the Python interpreter or an IDE, and type the following:

```
>>> import sys
>>> print sys.path
['', 'C:\\WINDOWS\\SYSTEM32\\python27.zip',
'C:\\Python27\\ArcGIS10.5\\Dlls', 'C:\\Python27\\ArcGIS10.5\\lib',
'C:\\Python27\\ArcGIS10.5\\lib\\plat-win', 'C:\\Python27\\ArcGIS10.5',
'C:\\Program Files (x86)\\ArcGIS\\Desktop10.5\\arcpy', 'C:\\Program Files
(x86)\\ArcGIS\\Desktop10.5\\ArcToolbox\\Scripts']
```

The system path (stored in the sys property `sys.path`) includes all of the folders that ArcPy requires to automate ArcGIS. The system path incorporates all directories listed in the `PYTHONPATH` environment variable (if one has been created); this is separate from the Windows Path environment variable discussed earlier. The two separate path variables work together to help Python locate modules.

The sys.path.append method

The `sys.path` property is a mutable list, and can be appended or extended to include new file paths that will point to modules the user wants to import. To avoid the need to adjust the `sys.path`, copy the module into the `site-packages` folder; however, this is not always possible, so use the `sys.path.append` method as needed:

```
>>> sys.path.append("C:\\Projects\\Requests")
>>> sys.path
['', 'C:\\WINDOWS\\SYSTEM32\\python27.zip',
'C:\\Python27\\ArcGIS10.5\\Dlls', 'C:\\Python27\\ArcGIS10.5\\lib',
'C:\\Python27\\ArcGIS10.5\\lib\\plat-win', 'C:\\Python27\\ArcGIS10.5',
'C:\\Program Files (x86)\\ArcGIS\\Desktop10.5\\arcpy', 'C:\\Program Files
(x86)\\ArcGIS\\Desktop10.5\\ArcToolbox\\Scripts','C:\\Projects\\Requests']
```

When the `sys.path.append` method is used, the adjustment is temporary. Adjust the `PYTHONPATH` environment variable in the Windows System Properties menu (discussed earlier in the Path environment variable section) to make a permanent change (and create the `PYTHONPATH` if it has not been created).

One last, valuable note: to import a module without adjusting the system path or copying the module into the `site-packages` folder, place the module in the folder that contains the script that is importing it. As long as the module is configured correctly, it will work normally. This is useful when there is no administrative access available to the executing machine.

Summary

In this chapter, we covered the basics of programming using Python, and introduced important Python modules. We covered how Python executes scripts and commands, and touched on the development environments used to craft scripts. In particular, we discussed Python basics including data types, containers and looping, how a Python script is executed by the Python interpreter, where the Python interpreter is located within the Python folder structure, and what the different Python script extensions mean (`.py`, `.pyc`, `.pyw`). We also covered Integrated Development Environments, and how they compare and contrast.

In the next chapter, we will explain how to use **ModelBuilder** to convert a modeled analysis into a Python script, and how to add more functionality to the exported script.

2
Creating the First Python Script

With the Python environment configured to fit our needs, we can now create and execute ArcPy scripts. To ease into the creation of Python scripts, this chapter will use ArcGIS **ModelBuilder** to model a simple analysis, and export it as a Python script. ModelBuilder is very useful for creating Python scripts. It has an operational and a visual component, and all models can be outputted as Python scripts, where they can be further customized.

This chapter we will cover the following topics:

- Modeling a simple analysis using ModelBuilder
- Exporting the model out to a Python script
- Windows file paths versus Pythonic file paths
- String formatting methods

Prerequisites

The following are the prerequisites for this chapter: ArcGIS 10x and Python 2.7, with arcpy available as a module.

For this chapter, the accompanying data and scripts should be downloaded from Packt Publishing's website. The completed scripts are available for comparison purposes, and the data will be used for this chapter's analysis. To run the code and test code examples, use your favorite IDE or open the **IDLE (Python GUI)** program from the `Start Menu/ArcGIS/Python2.7` folder after installing ArcGIS for Desktop. Use the built-in "interpreter" or code entry interface, indicated by the triple chevron >>> and a blinking cursor.

ModelBuilder

ArcGIS has been in development since the 1970s. Since that time, it has included a variety of programming languages and tools to help GIS users automate analysis and map production. These include the **Avenue** scripting language in the ArcGIS 3x series, and the **ARC Macro Language** (**AML**) in the ARCInfo Workstation days, as well as VBScript up until ArcGIS 10x, when Python was introduced. Another useful tool introduced in ArcGIS 9x was ModelBuilder, a visual programming environment used for both modeling analysis and creating tools that can be used repeatedly with different input feature classes.

A useful feature of ModelBuilder is an export function, which allows modelers to create Python scripts directly from a model. This makes it easier to compare how parameters in a ModelBuilder tool are accepted as compared to how a Python script calls the same tool and supplies its parameters, and how generated feature classes are named and placed within the file structure. ModelBuilder is a helpful tool on its own, and its Python export functionality makes it easy for a GIS analyst to generate and customize ArcPy scripts.

Creating a model and exporting to Python

This chapter and the associated scripts depend on the downloadable file `SanFrancisco.gdb` geodatabase available from Packt. `SanFrancisco.gdb` contains data downloaded from `https://datasf.org/` and the US Census' American Factfinder website at `https://factfinder.census.gov/faces/nav/jsf/pages/index.xhtml`. All census and geographic data included in the geodatabase is from the 2010 census. The data is contained within a feature dataset called **SanFrancisco**. The data in this feature dataset is in *NAD 83 California State Plane Zone 3*, and the linear unit of measure is the US foot. This corresponds to SRID **2227** in the **European Petroleum Survey Group** (**EPSG**) format.

The analysis which we will create with the model, and eventually export to Python for further refinement, will use bus stops along a specific line in San Francisco. These bus stops will be buffered to create a representative region around each bus stop. The buffered areas will then be intersected with census blocks to find out how many people live within each representative region around the bus stops.

Modeling the Select and Buffer tools

Using **ModelBuilder**, we will model the basic bus stop analysis. Once it has been modeled, it will be exported as an automatically generated Python script. Follow these steps to begin the analysis:

1. Open up ArcCatalog, and create a folder connection to the folder containing `SanFrancisco.gdb`. I have put the geodatabase in a `C` drive folder called "Projects" for a resulting file path of `C:\\Projects\\SanFrancisco.gdb`.

2. Right-click on **geodatabase,** and add a new toolbox called **Chapter2Tools**.

3. Right-click on **geodatabase;** select **New**, and then **Feature Dataset,** from the menu. A dialogue will appear that asks for a name; call it **Chapter2Results**, and push **Next**. It will ask for a spatial reference system; enter **2227** into the search bar, and push the magnifying glass icon. This will locate the correct spatial reference system: **NAD 1983 StatePlane California III FIPS 0403 Feet**. Don't select a vertical reference system, as we are not doing any Z value analysis. Push next, select the default tolerances, and push **Finish**.

4. Next, open ModelBuilder using the ModelBuilder icon or by right-clicking on the **Toolbox**, and create a new Model. Save the model in the Chapter2Tools toolbox as `Chapter2Model1`. Drag in the `Bus_Stops` feature class and the **Select** tool from the `Analysis/Extract` toolset in ArcToolbox. Open up the **Select** tool, and name the output feature class **Inbound71**. Make sure that the feature class is written to the `Chapter2Results` feature dataset. Open up the Expression SQL Query Builder, and create the following SQL expression : `NAME = '71 IB' AND BUS_SIGNAG = 'Ferry Plaza'`.

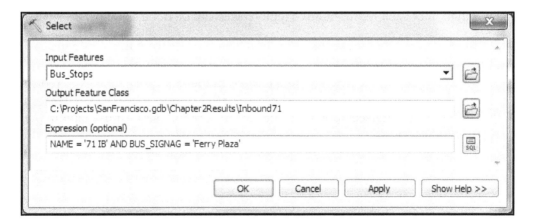

5. The next step is to add a Buffer Tool from the `Analysis/Proximity` toolset. The Buffer tool will be used to create buffers around each bus stop. The buffered bus stops allow us to intersect with census data in the form of census blocks, creating the representative regions around each bus stop.

6. Connect the output of the Select tool (**Inbound71**) to the Buffer tool. Open up the Buffer tool, add **400** to the **Distance** field, and change the units to **Feet**. Leave the rest of the options blank. Click on **OK,** and return to the model:

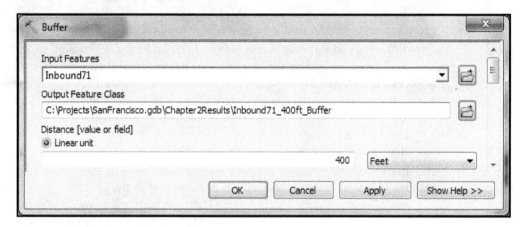

Adding in the Intersect tool

Now that we have selected the bus line of interest, and buffered the stops to create representative regions, we will need to intersect the regions with the census blocks to find the population of each representative region. This can be done as follows:

1. First, add the `CensusBlocks2010` feature class from the `SanFrancisco` feature dataset to the model.

2. Next, add in the Intersect tool located in the `Analysis/Overlay` toolset in the ArcToolbox. While we could use a **Spatial Join** to achieve a similar result, I have used the Intersect tool to capture the area of intersect for use later in the model and script.

At this point, our model should look like this:

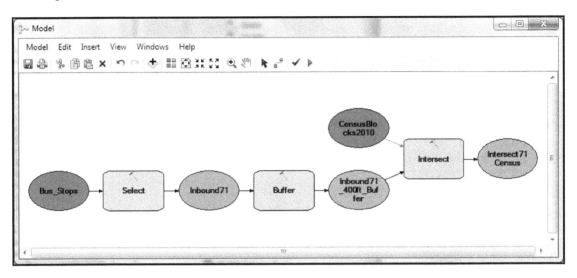

Tallying the analysis results

After we have created this simple analysis, the next step is to determine the results for each bus stop. Finding the number of people that live in census blocks, touched by the 400-foot buffer of each bus stop, involves examining each row of data in the final feature class, and selecting rows that correspond to the bus stop. Once these are selected, a sum of the selected rows would be calculated either using the `Field Calculator` or the `Summarize` tool. All of these methods will work, and yet none are perfect. They take too long, and worse, are not repeatable automatically if an assumption in the model is adjusted (if the buffer is adjusted from 400 feet to 500 feet, for instance).

This is where the traditional uses of ModelBuilder begin to fail analysts. It should be easy to instruct the model to select all rows associated with each bus stop, and then generate a summed population figure for each bus stop's representative region. It would be even better to have the model create a spreadsheet to contain the final results of the analysis. It's time to use Python to take this analysis to the next level.

Exporting the model and adjusting the script

While modeling analysis in ModelBuilder has its drawbacks, there is one fantastic option built into ModelBuilder: the ability to create a model, and then export the model to Python. Along with the ArcGIS Help Documentation, it is the best way to discover the correct Python syntax to use when writing ArcPy scripts.

Create a folder that can hold the exported scripts next to the SanFrancisco geodatabase (for example, C:\\ProjectsScripts). This will hold both the exported scripts that ArcGIS automatically generates, and the versions that we will build from those generated scripts. Now, perform the following steps:

1. Open up the model called Chapter2Model1.
2. Click on the **Model** menu in the upper-left side of the screen.
3. Select **Export** from the menu.
4. Select **To Python Script**.
5. Save the script as Chapter2Model1.py.

 Note that there is also the option to export the model as a graphic. Creating a graphic of the model is a good way to share what the model is doing with other analysts without the need to share the model and the data, and can also be useful when sharing Python scripts as well.

The Automatically generated script

Open the automatically generated script in an IDE. It should look like this:

```python
# -*- coding: utf-8 -*-
# ------------------------------------------------------------------
# Chapter2Model1.py
# Created on: 2017-01-26 04:26:31.00000
# (generated by ArcGIS/ModelBuilder)
# Description:
# ------------------------------------------------------------------

# Import arcpy module
import arcpy


# Local variables:
Bus_Stops = "C:\\Projects\\SanFrancisco.gdb\\SanFrancisco\\Bus_Stops"
CensusBlocks2010 =
```

```
    "C:\\Projects\\SanFrancisco.gdb\\SanFrancisco\\CensusBlocks2010"
Inbound71 =
    "C:\\Projects\\SanFrancisco.gdb\\Chapter2Results\\Inbound71"
Inbound71_400ft_Buffer =
"C:\\Projects\\SanFrancisco.gdb\\Chapter2Results\\Inbound71_400ft_Buffer"
Intersect71Census =
    "C:\\Projects\\SanFrancisco.gdb\\Chapter2Results\\Intersect71Census"


# Process: Select
arcpy.Select_analysis(Bus_Stops, Inbound71, "NAME = '71 IB' AND
    BUS_SIGNAG = 'Ferry Plaza'")


# Process: Buffer
arcpy.Buffer_analysis(Inbound71, Inbound71_400ft_buffer, "400 Feet",
    "FULL", "ROUND", "NONE", "")


# Process: Intersect
arcpy.Intersect_analysis([Inbound71_400ft_Buffer,CensusBlocks2010],
    Intersect71Census, "ALL", "", "INPUT")
```

Let's examine this script line by line. The first line is preceded by a pound sign ("#"), which again means that this line is a comment; however, it is not ignored by the Python interpreter when the script is executed as usual, but is used to help Python interpret the encoding of the script as described here: http://legacy.python.org/dev/peps/pep-0263.

The second commented line and the third line are included for decorative purposes. The next four lines, all commented, are used for providing readers information about the script: what it is called and when it was created along with a description, which is pulled from the model's properties. Another decorative line is included to visually separate out the informative header from the body of the script. While the commented information section is nice to include in a script for other users of the script, it is not necessary.

The body of the script, or the executable portion of the script, starts with the import arcpy line. Import statements are, by convention, included at the top of the body of the script. In this instance, the only module that is being imported is ArcPy.

ModelBuilder's export function creates not only an executable script, but also comments each section to help mark the different sections of the script. The comments let the user know where the variables are located, and where the **ArcToolbox** tools are being executed.

After the import statements come the variables. In this case, the variables represent the file paths to the input and output feature classes. The variable names are derived from the names of the feature classes (the base names of the file paths). The file paths are assigned to the variables using the assignment operator ("="), and the parts of the file paths are separated by two backslashes.

File paths in Python

To store and retrieve data, it is important to understand how file paths are used in Python as compared to how they are represented in Windows. In Python, file paths are strings, and strings in Python have special characters used to represent tabs "\t", newlines "\n", or carriage returns "\r", among many others. These special characters all incorporate single backslashes, making it very hard to create a file path that uses single backslashes. File paths in Windows Explorer all use single backslashes.

```
Windows Explorer:
C:\Projects\SanFrancisco.gdb\Chapter2Results\Intersect71Census
```

Python was developed within the Linux environment, where file paths have forward slashes. There are a number of methods used to avoid this issue. The first is using filepaths with forward slashes. The Python interpreter will understand file paths with forward slashes as seen in this code:

```
Python version:
"C:/Projects/SanFrancisco.gdb/Chapter2Results/Intersect71Census"
```

Within a Python script, the Python file path with the forward slashes will definitely work, while the Windows Explorer version might cause the script to throw an exception as Python strings can have special characters like the newline character \n, or tab \t. that will cause the string file path to be read incorrectly by the Python interpreter.

Another method used to avoid the issue with special characters is the one employed by ModelBuilder when it automatically creates the Python scripts from a model. In this case, the backslashes are "escaped" using a second backslash. The preceding script uses this second method to produce the following results:

```
Python escaped version:
"C:\\Projects\\SanFrancisco.gdb\\Chapter2Results\\Intersect71Census"
```

The third method, which I use when copying file paths from ArcCatalog or Windows Explorer into scripts, is to create what is known as a "raw" string. This is the same as a regular string, but it includes an "r" before the script begins. This "r" alerts the Python interpreter that the following script does not contain any special characters or escape characters. Here is an example of how it is used:

```
Python raw string:  r"C:\Projects\SanFrancisco.gdb\SanFrancisco\Bus_Stops"
```

Using raw strings makes it easier to grab a file path from Windows Explorer, and add it to a string inside a script. It also makes it easier to avoid accidentally forgetting to include a set of double backslashes in a file path, which happens all the time and is the cause of many script bugs.

String manipulation

There are three major methods for inserting variables into strings. Each has different advantages and disadvantages of a technical nature. It's good to know about all three, as they have uses beyond our needs here, so let's review them.

String manipulation method 1: string addition

String addition seems like an odd concept at first, as it would not seem possible to "add" strings together, unlike integers or floats which are numbers. However, within Python and other programming languages, this is a normal step. Using the plus sign "+", strings are "added" together to make longer strings, or to allow variables to be added into the middle of existing strings. Here are some examples of this process:

```
>>> aString = "This is a string"
>>> bString = " and this is another string"
>>> cString = aString + bString
>>> cString
```

The output is as follows:

```
'This is a string and this is another string'
```

Two or more strings can be "added" together, and the result can be assigned to a third variable for using it later in the script. This process can be useful for data processing and formatting.

Another similar offshoot of string addition is string multiplication, where strings are multiplied by an integer to produce repeating versions of the string, like this:

```
>>> "string" * 3
'stringstringstring'
```

String manipulation method 2: string formatting #1

The second method of string manipulation, known as string formatting, involves adding placeholders into the string, which accept specific kinds of data. This means that these special strings can accept other strings as well as integers and float values. These placeholders use the modulo "%" and a key letter to indicate the type of data to expect. Strings are represented using %s, floats using %f, and integers using %d. The floats can also be adjusted to limit the digits included by adding a modifying number after the modulo. If there is more than one placeholder in a string, the values are passed to the string in a tuple.

This method has become less popular, since the third method discussed next was introduced in Python 2.6, but it is still valuable to know, as many older scripts use it. Here is an example of this method:

```
>>> origString = "This string has as a placeholder %s"
>>> newString = origString % "and this text was added"
>>> print newString
```

The output is as follows:

```
This string has as a placeholder and this text was added
```

Here is an example when using a `float` placeholder:

```
>>> floatString1 = "This string has a float here: %f"
>>> newString = floatString % 1.0
>>> newString = floatString1 % 1.0
>>> print newString
```

The output is as follows:

```
This string has a float here: 1.000000
```

Here is another example when using a `float` placeholder:

```
>>> floatString2 = "This string has a float here: %.1f"
>>> newString2 = floatString2 % 1.0
>>> print newString2
```

The output is as follows:

```
This string has a float here: 1.0
```

Here is an example using an integer placeholder:

```
>>> intString = "Here is an integer: %d"
>>> newString = intString % 1
>>> print newString
```

The output is as follows:

```
Here is an integer: 1
```

String manipulation method 3: string formatting #2

The final method is known as string formatting. It is similar to the string formatting method #1, with the added benefit of not requiring a specific data type of placeholder. The placeholders, or tokens as they are also known, are only required to be in order to be accepted. The format function is built into strings; by adding `.format` to the string, and passing in parameters, the string accepts the values, as seen in the following example:

```
>>> formatString = "This string has 3 tokens: {0}, {1}, {2}"
>>> newString = formatString.format("String", 2.5, 4)
>>> print newString
This string has 3 tokens: String, 2.5, 4
```

The tokens don't have to be in order within the string, and can even be repeated by adding a token wherever it is needed within the template. The order of the values applied to the template is derived from the parameters supplied to the `.format` function, which passes the values to the string.

The third method has become my go-to method for string manipulation because of the ability to add the values repeatedly, and because it makes it possible to avoid supplying the wrong type of data to a specific placeholder, unlike the second method.

The ArcPy tools

After the import statements and the variable definitions, the next section of the script is where the analysis is executed. The same tools that we created in the model--the Select, Buffer, and Intersect tools--are included in this section. The same parameters that we supplied in the model are also included here: the inputs and outputs, plus the SQL statement in the Select tool, and the buffer distance in the Buffer tool.

The tool parameters are supplied to the tools in the script in the same order as they appear in the tool interfaces in the model. Here is the Select tool in the script:

```
arcpy.Select_analysis(Bus_Stops, Inbound71, "NAME = '71 IB' AND BUS_SIGNAG
= 'Ferry Plaza'")
```

It works like this: the arcpy module has a "method", or tool, called `Select_analysis`. This method, when called, requires three parameters: the input feature class (or shapefile), the output feature class, and the SQL statement. In this example, the input is represented by the variable `Bus_Stops`, and the output feature class is represented by the variable `Inbound71`, both of which are defined in the variable section. The SQL statement is included as the third parameter. Note that it could also be represented by a variable if the variable was defined before this line; the SQL statement, as a string, could be assigned to a variable, and the variable could replace the SQL statement as the third parameter. Here is an example of parameter replacement using a variable:

```
sqlStatement = "NAME = '71 IB' AND BUS_SIGNAG = 'Ferry Plaza'"
arcpy.Select_analysis(Bus_Stops, Inbound71, sqlStatement)
```

While ModelBuilder is good for assigning input and output feature classes to variables, it does not assign variables to every portion of the parameters. This will be an important thing to correct when we adjust and build our own scripts.

The Buffer tool accepts a similar set of parameters as the Select tool. There is an input feature class represented by a variable, an output feature class variable, and the distance that we provided (400 feet in this case) along with a series of parameters that were supplied by default. Note that the parameters rely on keywords, and these keywords can be adjusted within the text of the script to adjust the resulting buffer output. For instance, "Feet" could be adjusted to "Meters", and the buffer would be much larger. Check the help section of the tool to understand better how the other parameters will affect the buffer, and to find the keyword arguments that are accepted by the Buffer tool in ArcPy. Also, as noted earlier, all of the parameters could be assigned to variables, which can save time if the same parameters are used repeatedly throughout a script.

Sometimes, the supplied parameter is merely an empty string, as in this case here with the last parameter:

```
arcpy.Buffer_analysis(Inbound71,Inbound71_400ft_buffer,
                    "400 Feet", "FULL", "ROUND", "NONE", "")
```

The empty string for the last parameter, which, in this case, signifies that there is no dissolve field for this buffer, is found quite frequently within ArcPy. It could also be represented by two single quotes, but ModelBuilder has been built to use double quotes to encase strings.

The Intersect tool

The last tool, the Intersect tool, uses a different method to represent the files that need to be intersected together when the tool is executed. Because the tool accepts multiple files in the input section (meaning, there is no limit to the number of files that can be intersected together in one operation), it stores all of the file paths within one string. This string can be manipulated using one of the string manipulation methods discussed earlier, or it can be reorganized to accept a Python list that contains the file paths, or variables representing file paths as a list, as the first parameter in any order. The Intersect tool will find the intersection of all of the strings.

Adjusting the script

Now is the time to take the automatically generated script, and adjust it to fit our needs. We want the script to both produce the output data, and to have it analyze the data and tally the results into a spreadsheet. This spreadsheet will hold an averaged population value for each bus stop. The average will be derived from each census block that the buffered representative region surrounding the stops intersected.

Save the original script as "Chapter2Model1Modified.py".

Adding the CSV module to the script

For this script, we will use the csv module, a useful module for creating comma-separated salue spreadsheets. Its simple syntax will make it a useful tool for creating script outputs. ArcGIS for Desktop also installs the xlrd and xlwt modules, used to read or generate Excel spreadsheets respectively, when it is installed. These modules are also great for data analysis output.

After the `import arcpy` line, add `import csv`. This will allow us to use the `csv` module for creating the spreadsheet.

```
# Import arcpy module
import arcpy
import csv
```

The next adjustment is made to the Intersect tool. Notice that the two paths included in the input string are also defined as variables in the variable section. Remove the file paths from the input strings, and replace them with a list containing the variable names of the input datasets, as follows:

```
# Process: Intersect
arcpy.Intersect_analysis([Inbound71_400ft_buffer,CensusBlocks2010],Inte
    rsect71Census, "ALL", "", "INPUT")
```

Accessing the data: using a cursor

Now that the script is in place to generate the raw data we need, we need a way to access the data held in the output feature class from the Intersect tool. This access will allow us to aggregate the rows of data representing each bus stop. We also need a data container to hold the aggregated data in memory before it is written to the spreadsheet.

To accomplish the second part, we will use a Python dictionary. To accomplish the first part, we will use a method built into the ArcPy module: the Data Access SearchCursor.

The Python dictionary will be added after the Intersect tool. A dictionary in Python is created using curly brackets `{}`. Add the following line to the script, below the analysis section:

```
dataDictionary = {}
```

This script will use the bus stop IDs as keys for the dictionary. The values will be lists, which will hold all of the population values associated with each `busStopID`. Add the following lines to generate a Data Cursor:

```
with arcpy.da.SearchCursor(Intersect71Census, ["STOPID","POP10"]) as
    cursor:
        for row in cursor:
            busStopID = row[0]
            pop10 = row[1]
            if busStopID not in dataDictionary.keys():
                dataDictionary[busStopID] = [pop10]
            else:
                dataDictionary[busStopID].append(pop10)
```

This iteration combines a few ideas in Python and ArcPy. The `with...as` statement is used to create a variable (`cursor`), which represents the `arcpy.da.SearchCursor` object. It could also be written like this:

```
cursor = arcpy.da.SearchCursor(Intersect71Census, ["STOPID","POP10"])
```

The advantage of the `with...as` structure is that the `cursor` object is erased from memory when the iteration is completed, which eliminates locks on the feature classes being evaluated.

The `arcpy.da.SearchCursor` function requires an input feature class, and a list of fields to be returned. Optionally, an SQL statement can limit the number of rows returned.

The next line, `for row in cursor:`, is the iteration through the data. It is not a normal Pythonic iteration, a distinction that will have ramifications in certain instances. For instance, one cannot pass index parameters to the `cursor` object to only evaluate specific rows within the cursor object, as one can do with a list.

When using a Search Cursor, each row of data is returned as a tuple, which cannot be modified. The data can be accessed using indexes, as shown in the preceding code, where the two members of the tuple are assigned to variables.

The `if...else` condition allows the data to be sorted. As noted earlier, the bus stop ID, which is the first member of the data included in the tuple, will be used as a key. The conditional evaluates if the bus stop ID is included in the dictionary's existing keys (which are contained in a list, and accessed using the `dictionary.keys()` method). If it is not, it is added to the keys, and assigned a value that is a list that contains (at first) one piece of data, the population value contained in that row. If it does exist in the keys, the list is appended with the next population value associated with that bus stop ID. With this code, we have now sorted each census block population according to the bus stop with which it is associated.

Next we need to add code to create the spreadsheet. This code will use the same `with...as` structure, and will generate an average population value by using two built-in Python functions: `sum`, which creates a sum from a list of numbers, and `len`, which will get the length of a list, tuple, or string.

```
with open(r'C:\Projects\Averages.csv', 'wb') as csvfile:
    csvwriter = csv.writer(csvfile, delimiter=',')
    for busStopID in dataDictionary.keys():
        popList = dataDictionary[busStopID]
```

```
averagePop = sum(popList)/len(popList)
data = [busStopID, averagePop]
csvwriter.writerow(data)
```

The average population value is retrieved from the dictionary using the `busStopID` key, and then assigned to the variable `averagePop`. The two data pieces, the `busStopID` and the `averagePop` variable are then added to a list. This list is supplied to a `csvwriter` object, which knows how to accept the data and write it out to a file located at the file path supplied to the built-in Python function `open`, used to create simple files.

The script is complete, although it is nice to add one more line to the end to give us visual confirmation that the script has run.

```
print "Data Analysis Complete"
```

This last line will create an output indicating that the script has run. Once it is done, go to the location of the output CSV file and open it using Excel or Notepad, and see the results of the analysis. Our first script is complete!

Exceptions and tracebacks

During the process of writing and testing scripts, there will be errors that cause the code to break and throw exceptions. In Python, these are reported as a "traceback", which shows the last few lines of code executed before an exception occurred. To best understand the message, read them from the last line up. It will tell you the type of exception that occurred, and preceding to that will be the code that failed, with a line number, that should allow you to find and fix the code. It's not perfect, but it works.

Overwriting files

One common issue is that ArcGIS for Desktop does not allow you to overwrite files without turning on an environment variable. To avoid this issue, you can add a line after the import statements that will make overwriting files possible. **Be aware that the original data will be unrecoverable once it is overwritten.** It uses the `env` module to access the ArcGIS environment:

```
import arcpy
arcpy.env.overwriteOutput = True
```

The final script

Here is how the script should look in the end:

```
# Chapter2Model1Modified.py
# Import arcpy module
import arcpy
import csv

arcpy.env.overwriteOutput = True

# Local variables:
Bus_Stops = r"C:\Projects\SanFrancisco.gdb\SanFrancisco\Bus_Stops"
CensusBlocks2010 =
 r"C:\Projects\SanFrancisco.gdb\SanFrancisco\CensusBlocks2010"
Inbound71 = r"C:\Projects\SanFrancisco.gdb\Chapter2Results\Inbound71"
Inbound71_400ft_buffer =
 r"C:\Projects\SanFrancisco.gdb\Chapter2Results\Inbound71_400ft_buffer"
Intersect71Census =
 r"C:\Projects\SanFrancisco.gdb\Chapter2Results\Intersect71Census"

# Process: Select
arcpy.Select_analysis(Bus_Stops,
                      Inbound71,
                      "NAME = '71 IB' AND BUS_SIGNAG = 'Ferry Plaza'")

# Process: Buffer
arcpy.Buffer_analysis(Inbound71,
                      Inbound71_400ft_buffer,
                      "400 Feet", "FULL", "ROUND", "NONE", "")

# Process: Intersect
arcpy.Intersect_analysis([Inbound71_400ft_buffe,CensusBlocks2010],
                         Intersect71Census, "ALL", "", "INPUT")

dataDictionary = {}
with arcpy.da.SearchCursor(Intersect71Census, ["STOPID","POP10"]) as
  cursor:
    for row in cursor:
        busStopID = row[0]
        pop10 = row[1]
        if busStopID not in dataDictionary.keys():
            dataDictionary[busStopID] = [pop10]
        else:
            dataDictionary[busStopID].append(pop10)

with open(r'C:\Projects\Averages.csv', 'wb') as csvfile:
    csvwriter = csv.writer(csvfile, delimiter=',')
```

```
    for busStopID in dataDictionary.keys():
        popList = dataDictionary[busStopID]
        averagePop = sum(popList)/len(popList)
        data = [busStopID, averagePop]
        csvwriter.writerow(data)

print "Data Analysis Complete"
```

Summary

In this chapter, you learned how to craft a model of an analysis and export it out to a script. In particular, you learned how to use ModelBuilder to create an analysis and export it out as a script and how to adjust the script to be more "Pythonic". After explaining about the auto-generated script, we adjusted the script to include a results analysis and summation, which was outputted to a CSV file. We also briefly touched on the use of Search Cursors, which will be covered in greater detail in Chapter 3, *ArcPy Cursors: Search, Insert, and Update*. Also, we saw how built-in modules such as the csv module can be used along with ArcPy to capture analysis output in formatted spreadsheets.

In the next chapter, we will investigate the powerful data access module and its Search Cursors, Update Cursors, and Insert Cursors.

3
ArcPy Cursors - Search, Insert, and Update

Now that you understand how to interact with ArcToolbox tools using ArcPy, you will have a basic understanding of how to improve GIS work flows using Python. In this chapter, we will cover data cursors and the data access module introduced in ArcGIS 10.1.

Data cursors are used to access data records contained within data tables, using a row-by-row iterative approach. The concept was imported from relational databases, where data cursors are used to extract data from tables returned from an SQL expression. Cursors are used not only to search for data, but also to update data, or to add new data.

When we discuss creating data searches using ArcPy cursors, we are not just talking about attribute information. The new data access model cursors can interact directly with the shape field, and when combined with ArcPy Geometry objects, can perform geospatial functions, and replace the need to pass data to ArcToolbox tools. Data access cursors represent the most useful innovation yet in the realm of Python automation for GIS.

In this chapter, we will cover the following topics:

- Introduction to Python functions
- Using Search cursors to iterate through attribute and spatial data
- Using Update cursors to adjust values within rows
- Using Insert cursors to add new data to a dataset
- Using conditionals to update specific records

Python functions – avoid repeating code

Programming languages share a concept that has aided programmers for decades: functions. The idea of a function, loosely speaking, is to create blocks of code that will perform an action on a piece of data, transforming it as required by the programmer, and returning the transformed data back to the main body of the code script. We've already been introduced to some of Python's built-in functions in the last few chapters: the `int` function, for instance, will convert a string, or a floating number into an integer. Now it's time to write our own.

Functions are used because they solve many different needs within programming. Functions reduce the need to write repetitive code, which, in turn, reduces the time needed to create a script. They can be used to create ranges of numbers (the `range` function), or to determine the maximum value of a list (the `max` function), or to create an SQL statement to select a set of rows from a feature class (see later). They can even be copied and used in another script, or included as part of a module that can be imported into scripts. Function reuse has the added bonus of making programming more useful, and less of a chore. When a script writer starts writing functions, it is a major step towards making programming part of a GIS workflow.

Technical definition of functions

Functions, also called subroutines or procedures in other programming languages, are special blocks of code that have been designed to either accept input data and transform it, or to provide data to the main program when called without any input required. In theory, the functions should only transform data that has been provided to the function as a parameter; it should not change any other part of the script that has not been included in the function. To make this possible, the concept of namespaces is invoked. Namespaces are used to isolate variables within a script: variables are either global, and available to be used in the main body of a script as well as in a function, or they are local, and only available within a function.

Namespaces make it possible to use a variable name within a function, and allow it to represent a value while also using the same variable name in another part of the script. This becomes especially important when importing modules from other programmers; within that module and its functions, the variables that it contains might have a variable name that is the same as a variable name within the main script.

In a high-level programming language such as Python, there is built-in support for functions including the ability to define function names and the data inputs (known as parameters). Functions are created using the keyword def plus a function name along with parentheses that may or may not contain parameters--the expected input data values. Parameters can also be defined with default values, so the parameters only need to be passed to the function when they differ from the default, as needed. The values that are returned from the function are also easily defined.

The first function

Let's create a function to get a feel of what is possible when writing functions. First, we need to invoke the function by providing the def keyword and providing a name along with the parentheses. The function called firstFunction will return a string when called, using the return keyword. An optional "doc string" is added below the name of the function; this is used for documentation generation and when reading code to understand what each function does:

```
def firstFunction():
    'a simple function returning a string'
    stringvariable = "My First Function"
    return stringvariable

>>> firstFunction()
```

The output is as follows:

```
'My First Function'
```

The function is defined and given a name, and then the parentheses are added, followed by a colon. The following lines must then be indented (a good IDE will add the indention automatically). The function does not have any parameters, so the parentheses are empty. The function then uses the keyword return to return a value, in this case a string, from the function.

Next, the function is called by adding parentheses to the function name. When it is called, it will do what it has been instructed to do: return the string.

Functions with parameters

Now let's create a function that accepts parameters and transforms them as needed. This function will accept a number and multiply it by three as shown:

```
def secondFunction(number):
    'this function multiples numbers by 3'
    return number *3
>>> secondFunction(4)
```

The output is as follows:

```
12
```

Note that this will return a float if a float is passed to it:

```
>>> secondFunction(4.0)
```

The output is as follows:

```
12.0
```

This function can even be used on strings, as they can be multiplied in their own unique way:

```
>>> secondFunction("String")
```

The output is as follows:

```
StringStringString
```

The function has one flaw, however: we will have to redesign it to only work on numbers, as there is no assurance that the value passed to the function is a number. We need to add a condition to the function to make sure it does not throw an exception:

```
def secondFunction(number):
    'this function multiples numbers by 3'
    if type(number) == type(1) or type(number) == type(1.0):
        return number *3
>>> secondFunction(4.0)
```

The output is as follows:

```
12.0
```

The function now accepts a parameter, checks what type of data it is, and returns a multiple of the parameter if it is an integer or a function. If it is a string or some other data type, as in the last example, no value is returned.

There is one more adjustment to the simple function that we should discuss: parameter defaults. By including default values in the definition of the function, we can avoid having to provide parameters that rarely change. If, for instance, we want a different multiplier than 3 in the simple function, we would define it with a default parameter like this:

```
def thirdFunction(number, multiplier=3):
    'this function multiples numbers by 3'
    if type(number) == type(1) or type(number) == type(1.0):
        return number *multiplier

>>> thirdFunction(4)
```

The output is as follows:

```
12
```

If we pass another value in the second position, it will override the default parameter.

```
>>> thirdFunction(4,5)
```

The output is as follows:

```
20
```

These simple functions combine many of the concepts that we discussed in earlier chapters including built-in functions, conditionals, parameters, parameter defaults, and function returns. We can now move on to creating functions with ArcPy.

Using functions to replace repetitive code

One of the main uses of functions is to ensure that the same code does not have to be written over and over. Let's return to our example from the last chapter, and make a function from the last portion of the script. Because we will need to create CSV files often, it is best to convert the CSV processing into a custom Python function.

The createCSV function

The `createCSV` function accepts the dictionary and the name of the output CSV file as strings. It also uses a parameter with a default value, `mode`, which can be adjusted to either create or add to an existing file (`ab`), or to overwrite an existing file (`wb`).

```
def createCSV(data, csvname, mode ='ab'):
    with open(csvname, mode) as csvfile:
        csvwriter = csv.writer(csvfile, delimiter=',')
        csvwriter.writerow(data)
```

Creating an XLS using XLWT

XLWT is a module that can generate Microsoft Excel spreadsheets just as easily as we have been generating CSV files, and has a multitude of styling options. We can create a reusable function for accepting data that will generate a workbook, add a sheet, add data to the sheet, and save the sheet to a file location

This function requires three parameters: dataset, a list containing rows of data in list format; a sheet name as a string, and a string file name that ends with the `.xls` extension. It uses the enumerate function to count each iteration through the rows of data, and an enumerate function to count each item in the row; these two counters are used to place the item in the correct spreadsheet cell.

```
def generateXLS(dataset, sheetName, fileName):
    import xlwt
    workbook = xlwt.Workbook()
    sheet = workbook.add_sheet(sheetName)
    for YCOUNTER, data in enumerate(dataset):
        for XCOUNTER, value in enumerate(data):
            sheet.write(YCOUNTER, XCOUNTER, value)
    workbook.save(fileName)
```

XLWT and XLRD are both installed with Python 2.7 when ArcGIS for Desktop is installed. XLRD is used to read spreadsheets, while XLWT is used to write them. Explore their extensive documentation here:

XLWT: `http://xlwt.readthedocs.io/en/latest/`

XLRD: `http://xlrd.readthedocs.io/en/latest/`

The data access module

Introduced with the release of ArcGIS 10.1, the new data access module known as `arcpy.da` has made data interaction easier and faster than allowed by previous data cursors. By allowing for direct access to the shape field in a variety of forms (shape object, *X* values, *Y* values, centroid, area, length, and more), and a variety of formats (**JavaScript Object Notation (JSON)**, **Keyhole Markup Language (KML)**, **Well-Known Binary (WKB)**, and **Well-Known Text (WKT)**), the data access module greatly increases the ability of a GIS analyst to extract and control shape field data.

The data access cursors accept a number of required and optional parameters. The required parameters are the path to the feature class as a string (or a variable representing the path) and the fields to be returned. If all fields are desired, use the asterisk notation, and provide a list with an asterisk as a string as the fields parameter (["*"]). If only a few fields are required, provide those fields as string field names (for example, ["NAME", "DATE"]).

The other parameters are optional, but are very important for both search and Update cursors. A `where` clause, in the form of an SQL expression, can be provided next; this clause will limit the number of rows returned from the dataset (as demonstrated by the SQL expression in the scripts in the last chapter). The SQL expressions used by the search and Update cursors are not complete SQL expressions, as the `SELECT` or `UPDATE` commands are provided automatically by the choice of cursor. Only the `where` clause of the SQL expression is required for this parameter.

A spatial reference can be provided next in the ArcPy spatial reference format; this is not necessary if the data is in the correct format, but can be used to transform data into another projection on the fly. There is no way to specify the spatial transformation used, however. The third optional parameter is a Boolean (or True/False) value, which declares if data should be returned as exploded points (that is, a list of the individual vertices) or in the original geometry format. The final optional parameter is another list, which can be used to organize the data returned by the cursor; this list would include SQL keywords such as `DISTINCT`, `ORDER BY`, or `GROUP BY`. However, this final parameter is only available when working with a geodatabase.

Search cursors

The `arcpy.da.SearchCursor` method is used to retrieve data from datasets. Both the shape data and the attribute data in a feature class or shapefile can be accessed using this method (however, data cannot be updated or created). The required method parameters are a feature class or shapefile file path, and a list of fields to return. An optional SQL where conditional can be passed, is to limit the number of data rows returned.

To access the data rows, the `arcpy.da.SearchCursor` is called by passing the required and optional parameters. The method returns a cursor object, which can be iterated using a for loop. During each loop, the cursor object provides one row of data selected from the dataset, and the items in the row can be accessed using indexing. The first item in a row is at index 0, the next at index 1, and so on. The order of the data items in each row corresponds to the order of the fields list passed to the method.

Let's take a look at using `arcpy.da.SearchCursor` for shape field interactions. If we need to produce a spreadsheet listing all the bus stops along a particular route, and include the location of the data in an *X/Y* format, we could use the Add *XY* tool from the ArcToolbox. However, this has the effect of adding two new fields to our data, which is not always allowed, especially when the data is stored in enterprise geodatabases with fixed schemas. Instead, we'll use the `SHAPE@XY` token built into the data access module to easily extract the data, and pass it to the `createCSV` function described earlier along with the SQL expression limiting results to the stops of interest, as follows:

```
csvname = r"C:\Projects\StationLocations.csv"
headers = 'Bus Line Name','Bus Stop ID', 'X','Y'
createCSV(headers, csvname, 'wb')
sql = "NAME = '71 IB' AND BUS_SIGNAG = 'Ferry Plaza'"
with arcpy.da.SearchCursor(Bus_Stops,['NAME', 'STOPID', 'SHAPE@XY'],
  sql) as cursor:
    for row in cursor:
        linename = row[0]
        stopid = row[1]
        locationX = row[2][0]
        locationY = row[2][1]
        data = linename, stopid, locationX, locationY
        createCSV(data, csvname)
```

Note that each row of data returned by our search is a tuple, as the Search cursor does not allow any data manipulation, and tuples are immutable as soon as they are created. In contrast, data returned from Update cursors is in the list format, as lists can be updated. Both can be accessed using the indexing as shown earlier.

Each row returned by the cursor is a tuple with three objects: the name of the bus stop, the bus stop ID, and finally, another tuple containing the X/Y location of the stop. The objects in the tuple, contained in the variable `row`, are accessible using indexing: the bus stop name is at index 0, the ID is at index 1, and the location tuple is at index 2.

Within the location tuple, the *X* value is at index 0, and the *Y* value is at index 1; this makes it easy to access the data in the location tuple by passing a value as shown next:

```
locationX = row[2][0]
locationY = row[2][1]
```

The ability to add lists and tuples, and even dictionaries, to another list, tuple, or dictionary is a strong component of Python, and makes data access logical, and data organization easy.

However, the spreadsheet returned from the preceding code has a few issues: the location is returned in the native projection of the feature class (in this case, a State Plane projection), and there are rows of data that are repeated. It would be much more helpful if we could provide latitude and longitude values in the spreadsheet, and if the duplicate values were removed. Let's use the optional spatial reference parameter and a list to sort the data before we pass it to the `createCSV` function.

```
spatialReference = arcpy.SpatialReference(4326)
sql = "NAME = '71 IB' AND BUS_SIGNAG = 'Ferry Plaza'"
dataList = []
with arcpy.da.SearchCursor(Bus_Stops, ['NAME','STOPID','SHAPE@XY'],
  sql, spatialReference) as cursor:
    for row in cursor:
        linename = row[0]
        stopid = row[1]
        locationX = row[2][0]
        locationY = row[2][1]
        data = linename, stopid, locationX, locationY
        if data not in dataList:
            dataList.append(data)

csvname = "C:\Projects\StationLocations.csv"
headers = 'Bus Line Name','Bus Stop ID', 'X','Y'
createCSV(headers, csvname, 'wb')
for data in dataList:
    createCSV(data, csvname)
```

The spatial reference is created by passing a code representing the desired projection system. In this case, the code for the *WGS 1984* Latitude and Longitude geographic system is **4326**, and is passed to the `arcpy.SpatialReference` method to create a spatial reference object that can be passed to the Search cursor. Also, the `if` conditional is used to filter the data, accepting only one list per stop into the list called **dataList**. This new version of the code will produce a CSV file with the desired data.

Attribute field interactions

Apart from the shape field interactions, another improvement offered by the data access module cursors is the ability to pass specific fields as a parameter. Earlier, data cursors required the use of a less efficient "get value" function call, or required the fields to be called as if they were methods available to the function. The new method allows for all fields to be called by passing an asterisk, a valuable method to access fields in feature classes that have not been inspected previously.

One of the more valuable improvements is the ability to access the **unique ID** field without needing to know if the data set is a feature class or a shapefile. Because shapefiles had a feature ID or FID, and feature classes had an object ID, it was harder to program a script tool to access the unique ID field. Data access module cursors allow for the use of the `OID@` token to request the unique ID from either type of input. This makes the need to know the type of unique ID irrelevant.

As demonstrated earlier, other attribute fields are requested by string in a list. The field names must match the true name of the field; alias names cannot be passed to the cursor. The fields can be in the list in any order desired, and will be returned in the order requested. Only the required fields have to be included in the list.

Here is a demonstration of requesting field information:

```
sql = "OBJECTID = 1"
with arcpy.da.SearchCursor(Bus_Stops,['STOPID','NAME', 'OID@'], sql) as
  cursor:
    for row in cursor:
        data = row
>>> print data
(1111665, u'14 OB', 1)
```

If the fields in the fields list were adjusted, the data in the resulting row would reflect the adjustment. Also, all of the members of the tuple returned by the cursor are accessible by zero-based indexing.

Update cursors

Update cursors are used to adjust data within existing rows of data. Updates become very important when calculating data, or when converting null values to non-null ones. Combined with specific SQL expressions, data can be targeted for updating with newly collected or calculated values.

Note that running code containing an Update cursor will change, or update, the data on which it operates. It is a good idea to make a copy of the data to test out the code before running it on the original data.

All data access module Search cursor parameters discussed earlier are valid for Update cursors. The main difference is that data rows returned by Update cursors are returned as lists. Because lists are mutable, they can be adjusted using list value assignment.

As an example, let's imagine that the bus line 71 will be renamed as 75. Both inbound and outbound lines will be affected, so, an SQL expression must be included to get all rows of data associated with the line. Once the data cursor is created, the rows returned must have the name adjusted, added back into the list, and the Update cursor's updateRow method must be invoked. Here is how this scenario would look in code:

```
sql = "NAME LIKE '71%'"
with arcpy.da.UpdateCursor(Bus_Stops, ['NAME'],sql),) as cursor:
    for row in cursor:
        lineName = row[0]
        newName = lineName.replace('71','75')
        row[0] = newName
        cursor.updateRow(row)
```

The SQL expression will return all rows of data with a name starting with '71'; this will include '71 IB' and '71 OB'. Note that the SQL expression must be enclosed in double quotes, as the attribute value needs to be in single quotes.

For each row of data, the name at position zero in the row returned is assigned to the variable lineName. This variable, a string, uses the replace method to replace the characters '71' with the characters '75'. This could also just be replacing '1' with '5', but I wanted to be explicit as to what is being replaced.

Once the new string has been generated, it is assigned to the variable newName. This variable is then added to the list returned by the cursor using list assignment; this will replace the data value that initially occupied the zero position in the list. Once the row value has been assigned, it is then passed to the cursor's updateRow method. This method accepts the row, and updates the value in the feature class for that particular row.

Updating the shape field

For each row, all values included in the list returned by the cursor are available for update except the unique ID (while no exception will be thrown, the UID values will not be updated). Even the shape field can be adjusted with a few caveats. The main caveat is that the updated shape field must be the same geometry type as the original row: a point can be replaced with a point, a line with a line, and a polygon with another polygon.

Adjusting a point location

If a bus stop was moved down the street from its current position, it would need to be updated using an Update cursor. This operation will require a new location in the X/Y format, preferably in the same projection as the feature class to avoid any loss of location fidelity in a spatial transformation. There are two methods available to us for creating a new point location, depending on the method used to access the data. The first method is used when the location data is requested using the SHAPE@ tokens, and requires the use of an ArcPy Geometry type--in this case, the Point type. The ArcPy Geometry types are discussed in detail in the next chapter.

```
sql = 'OBJECTID < 5'
with arcpy.da.UpdateCursor(Bus_Stops, [ 'OID@', 'SHAPE@'],sql) as
   cursor:
     for row in cursor:
         row[1] = arcpy.Point(5999783.78657, 2088532.563956)
         cursor.updateRow(row)
```

By passing an X and Y value to the ArcPy Point geometry, a Point shape object is created and passed to the cursor in the updated list returned by the cursor. Assigning a new location to the shape field in a tuple, then using the cursor's updateRow method allows the shape field value to be adjusted to the new location. Because the first four bus stops are at the same location, they are all moved to the new location.

The second method applies to all other forms of shape field interactions including the SHAPE@XY, SHAPE@JSON, SHAPE@KML, SHAPE@WKT, and SHAPE@WKB tokens. These are updated by passing the new location in the format requested back to the cursor, and updating the list:

```
sql = 'OBJECTID < 5'
with arcpy.da.UpdateCursor(Bus_Stops, [ 'OID@', 'SHAPE@XY'],sql) as
   cursor:
     for row in cursor:
         row[1] =(5999783.786500007, 2088532.5639999956)
         cursor.updateRow(row)
```

Here is the same code using the `SHAPE@JSON` keyword and a JSON representation of the data:

```
sql = 'OBJECTID < 5'
with arcpy.da.UpdateCursor(Bus_Stops, [ 'OID@', 'SHAPE@JSON'],sql) as
   cursor:
      for row in cursor:
          print row
          row[1] = u'{"x":5999783.7865000069, "y":2088532.5639999956,
                    "spatialReference":{"wkid":102643}}'
          cursor.updateRow(row)
```

As long as the keyword, the data format, and the geometry type match, the location is updated to the new coordinates. The keyword method is very useful when updating points; however, the `SHAPE@XY` keyword does not work with lines or polygons, as the location returned represents the centroid of the requested geometry.

Deleting a row using an Update cursor

If we need to remove a row of data, the Update cursor has a `deleteRow` method, which works to remove the row. **Note that this will completely remove the data row, making it unrecoverable.** The `deleteRow` method does not require a parameter to be passed to it; instead, it removes the current row:

```
sql = 'OBJECTID < 2'
Bus_Stops = r'C:\Projects\SanFrancisco.gdb\SanFrancisco\Bus_Stops'
with arcpy.da.UpdateCursor(Bus_Stops, ['OID@', 'SHAPE@XY'],sql) as
   cursor:
      for row in cursor:
          cursor.deleteRow()
```

Using an Insert cursor

Now that we have a grasp on how to update existing data, let's investigate using Insert cursors to create new data, and add it to a feature class. The methods involved are very similar to using other data access cursors, except that we do not need to create an iterable cursor to extract rows of data; instead, we will create a cursor that will have the special `insertRow` method capable of adding data to the feature class row by row.

The Insert cursor can be called using the same `with...as` syntax, but generally, it is created as a variable in the flow of the script.

Note that only one cursor can be invoked at a time; an exception (a Python error) will be generated when creating two insert (or update) cursors without first removing the initial cursor--the Python `del` keyword is used to remove the cursor variable from memory. This is why the `with...as` syntax is preferred by many.

The data access module's Insert cursor requires some of the same parameters as the other cursors. The feature class to be written to and the list of fields that will have data inserted (this includes the shape field) are required. Spatial reference will not be used, as the new shape data must be in the same spatial reference as the feature class. No SQL expression is allowed for an Insert cursor.

The data to be added to the feature class will be in the form of a tuple or a list in the same order as the fields that are listed in the fields list parameter. Only fields of interest need to be included in the list of fields, meaning not every field needs a value in the list to be added. When adding a new row of data to a feature class, the unique ID will be automatically generated, making it unnecessary to explicitly include the unique ID (in the form of the `OID@` keyword) in the list of fields to be added.

Let's explore code that could be used to generate a new bus stop. We'll write to a test dataset called `TestBusStops`. We are only interested in the name and stop ID fields, so those fields, along with the shape field (which is in a State Plane projection system), will be included in the data list to be added.

```
Bus_Stops = r'C:ProjectsSanFrancisco.gdbTestBusStops'
insertCursor = arcpy.da.InsertCursor(Bus_Stops,
   ['SHAPE@','NAME','STOPID'])
coordinatePair = (6001672.5869999975, 2091447.0435000062)
newPoint = arcpy.Point(*coordinatePair)
dataList = [newPoint,'NewStop1',112121]
insertCursor.insertRow(dataList)
del insertCursor
```

If there is an iterable list of data to be inserted into the feature class, create the Insert cursor variable before entering the iteration, and delete the Insert cursor variable once the data has been iterated through. Alternatively, you can use the "with...as" method to automatically delete the Insert cursor variable when the iteration is complete, as seen in this code:

```
Bus_Stops = r'C:ProjectsSanFrancisco.gdbTestBusStops'
listOfLists = [[(6002672.58675, 2092447.04362),'NewStop2',112122],
               [(6003672.58675, 2093447.04362),'NewStop3',112123],
               [(6004672.58675, 2094447.04362),'NewStop4',112124]
               ]
with arcpy.da.InsertCursor(Bus_Stops, ['SHAPE@','NAME','STOPID']) as
   iCursor:
```

```
for dataList in listOfLists:
    newPoint = arcpy.Point(*dataList[0])
    dataList[0] = newPoint
    iCursor.insertRow(dataList)
```

As a list, the `listOfLists` variable is iterable. Each list within it is considered as `dataList` in the iteration, and the first value in `dataList`, the coordinate pair, is passed to the `arcpy.Point` function to create a `Point` object. The `arcpy.Point` function requires two parameters, *X* and *Y*; these are extracted from the coordinate pair tuple using the asterisk, which "explodes" the tuple, and passes the values it contains to the function. The `Point` object is then added back into `dataList` using index-based list assignment, which would not be available to us if the `dataList` variable was a tuple (we would instead have to create a new list, and add in the `Point` object and the other data values).

Inserting a polyline geometry

To create and insert a polyline type shape field from a series of points, it's best to use the `SHAPE@` keyword. We will also further explore the ArcPy geometry types, which will be discussed in the next chapter. When working with the `SHAPE@` keyword, we have to work with data in ESRI's spatial binary formats, and the data must be written back to the field in the same format using the ArcPy geometry types.

To create a polyline object, there is one required parameter: at least two valid ArcPy Point objects, made of two coordinate pairs, in an Array object. When working with the `SHAPE@` keyword, convert the coordinate pairs into an ArcPy point, then pass at least two points to an ArcPy array, and then pass the array to an ArcPy polyline. This polyline geometry will be inserted into the shape field of the feature class.

```
listOfPoints = [(6002672.58675, 2092447.04362),
                (6003672.58675, 2093447.04362),
                (6004672.58675, 2094447.04362)
                ]
line = 'New Bus Line'
lineID = 12345
busLine = r'C:\Projects\SanFrancisco.gdb\TestBusLine'
insertCursor = arcpy.da.InsertCursor(busLine, ['SHAPE@', 'LINE', 'LINEID'])
lineArray = arcpy.Array()
for pointsPair in listOfPoints:
    newPoint = arcpy.Point(*pointsPair)
    lineArray.add(newPoint)
newLine = arcpy.Polyline(lineArray)
insertData = newLine, line, lineID
insertCursor.insertRow(insertData)
```

The three coordinate pairs in tuples are iterated and converted into `Point` objects, which are, in turn, added to the Array object called `lineArray`. The `Array` object is then added to the `Polyline` object called `newLine`, which is then added to a tuple with the other data attributes, and inserted into the feature class by the Insert cursor.

Inserting a polygon geometry

Polygons are also inserted, or updated, using cursors. The ArcPy polygon geometry type does not require the coordinate pairs to include the first point twice (that is, as the first point and as the last point). The polygon is closed automatically by the `arcpy.Polygon` function. The array passed to the Polygon geometry must have at least three Points.

```
listOfPoints = [(6002672.58675, 2092447.04362), (6003672.58675,
    2093447.04362), (6004672.58675, 2093447.04362), (6004672.58675,
    2091447.04362)]
polyName = 'New Polygon'
polyID = 54321
blockPoly = r'C:\Projects\SanFrancisco.gdb\Chapter3Results\TestPolygon'
insertCursor = arcpy.da.InsertCursor(blockPoly,
    ['SHAPE@','BLOCK','BLOCKID'])
polyArray = arcpy.Array()
for pointsPair in listOfPoints:
    newPoint = arcpy.Point(*pointsPair)
    polyArray.add(newPoint)
newPoly = arcpy.Polygon(polyArray)
insertData = newPoly, polyName, polyID
insertCursor.insertRow(insertData)
del insertCursor
```

Here is a visualization of the result of the insert operation:

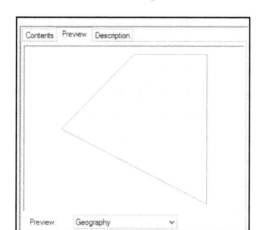

Summary

In this chapter, we covered the basic uses of data access module cursors. Search, Update, and Insert cursors were explored and demonstrated, and a special focus was placed on the use of these cursors for extracting shape data from the shape field. Cursor parameters were also introduced, including the spatial reference parameter and the SQL expression where clause parameter.

In the next chapter, we will further explore the use of cursors, especially with the use of ArcPy geometry types.

4
ArcPy Geometry Objects and Cursors

The essence of geospatial analysis is using geometric shapes--points, lines, and polygons--to model the geography of real-world objects and their location-based relationships. The simple shapes and their geometric properties of location, length, and area are processed using geospatial operations to generate analysis results. It is the combination of modeled geographic data and the associated attribute information that separates geospatial information systems from all other information systems.

Until ArcPy, processing feature class geometry using geospatial operations depended on the pre-built tools within ArcToolbox. ArcPy has made it possible to directly access the geometric shapes which are stored as mathematical representations in the shape field of feature classes. Once accessed, this geometric data is loaded into ArcPy geometry objects to make the data available for analysis within an ArcPy script. Because of this advance, writing scripts that access geometry fields and use them to perform analysis has transformed ArcGIS geospatial analysis. In this chapter, we'll explore how to generate and use the ArcPy geometry objects to perform geospatial operations, and apply them to the bus stops analysis.

In this chapter we will cover the following:

- `Point` and `Array` "constructor" objects
- `PointGeometry`, `Polyline`, and `Polygon` geometry objects
- Using the geometry objects to perform geospatial operations
- Integrating the geometry objects into scripts
- Performing common geospatial operations using the geometry objects
- Replacing the use of ArcToolbox tools in the script with geometry object methods

ArcPy geometry object classes

In designing geometry objects, the authors of ArcPy made it possible to perform geospatial operations in memory, reducing the need to use tools in the ArcToolbox for these operations. This results in speed gains, as there is no need to write the results of the calculations to disk at each step of the analysis. Instead, the results of the steps can be passed from function to function within the script. The final results of the analysis can be written to the hard drive as a feature class, or they can written onto a spreadsheet, or passed to another program.

The geometry objects are written as Python classes: special blocks of code that contain internal functions. The internal functions are the methods and properties of the geometry objects; when called, they allow the object to perform an operation (a *method*), or to reveal information about the geometry object (a *property*). Python classes are written with a `main` class that contains shared methods and properties, and with subclasses that reference the `main` class, but also have specific methods, and properties that are not shared. The `main` class is the ArcPy `Geometry` object, while the subclasses are the `PointGeometry`, `MultiPoint`, `Polyline`, and `Polygon` objects.

The geometry objects are generated in three ways. The first requires data cursors to read the existing feature classes, and pass a special keyword as a field name. The shape data returned by the cursor is a geometry object. The second method is to create new data by passing raw coordinates to a "constructor" object (either a `Point` or `Array` object), which is then passed to a geometry object. The third method is to read data from a feature class using the **Copy Features** tool from the ArcToolbox.

Each geometry object has methods that allow for read access and write access. The read access methods are important for accessing the coordinate points, which constitute the points, lines, and polygons. The write access methods are important when generating new data objects, which can be analyzed or written to disk.

The `PointGeometry`, `Multipoint`, `Polyline`, and `Polygon` geometry objects are used for performing analysis upon their respective geometry types. The generic geometry object can accept any geometry type and an optional spatial reference to perform geospatial operations when there is no need to discern the geometry type.

Two other ArcPy classes will be used for performing geospatial operations in memory: the `Array` object and the `Point` object. They are "constructor" objects, as they are not sub-classed from the geometry class, but are, instead, used to "construct" the geometry objects. The `Point` object is used to create coordinate points from raw coordinates. The `Array` object is a list of coordinate points that can be passed to a `Polyline` or `Polygon` object, as a regular Python list of ArcPy `Point` objects cannot be used to generate those geometry objects.

ArcPy Point objects

`Point` objects are the "building blocks" used to generate geometry objects. Also, all of the geometry objects will return component coordinates as `Point` objects when using read access methods. Point objects allow for simple geometry access using its **X**, **Y**, and **Z** properties, and a limited number of geospatial methods, such as `contains`, `overlaps`, `within`, `touches`, `crosses`, `equals`, and `disjoint`. Let's use IDLE to explore some of these methods with two `Point` geometry objects with the same coordinates, as follows:

```
>>> point   = arcpy.Point(4,5)
>>> point1  = arcpy.Point(4,5)
>>> point.equals(point1)
True
>>> point.contains(point1)
True
>>> point.crosses(point1)
False
>>> point.overlaps(point1)
False
>>> point.disjoint(point1)
False
>>> point.within(point1)
True
>>> point.X, point.Y
(4.0, 5.0)
```

In these preceding examples, we see some of the idiosyncrasies of the `Point` object. With two points that have the same coordinates, the results of the `equals` method and the `disjoint` method are as expected. The `disjoint` method will return `True` when the two objects do not share coordinates, while the opposite is true with the `equals` method. The `contains` method will work with the two `Point` objects, and return `True`. The `crosses` method and `overlaps` method produce somewhat surprising results, as the two `Point` objects do overlap in location, and could be considered to cross; however, these methods do not return the expected result, as they are not built to compare two points.

> There are a number of other methods available to `Point` objects. More information is available here:
>
> http://desktop.arcgis.com/en/arcmap/latest/analyze/arcpy-class
> es/point.htm

ArcPy Array objects

Before we progress to `Polyline` and `Polygon` objects, we need to understand the ArcPy `Array` object. It is the bridge between the `Point` objects and those geometry objects that require multiple coordinate points. `Array` objects accept `Point` objects as parameters, and the `Array` object is, in turn, passed as a parameter to the geometry object to be created. Let's use `Point` objects with an `Array` object to understand better how they work together.

The `Array` object is similar to a Python list, with `extend`, `append`, and `replace` methods, and also has unique methods such as `add` and `clone`. The `add` method will be used to add `Point` objects individually, as follows:

```
>>> point  = arcpy.Point(4,5)
>>> point1 = arcpy.Point(7,9)
>>> array = arcpy.Array()
>>> array.add(point)
>>> array.add(point1)
```

The `extend` method adds a list of `Point` objects all at once, instead of adding `Point` objects one at a time using `add`:

```
>>> point  = arcpy.Point(4,5)
>>> point1 = arcpy.Point(7,9)
>>> pList = [point,point1]
>>> array = arcpy.Array()
>>> array.extend(pList)
```

The `insert` method will put a `Point` object in the array at a specific index, while the `replace` method is used to replace a `Point` object in an array by passing an index and a new `Point` object like this:

```
>>> point   = arcpy.Point(4,5)
>>> point1  = arcpy.Point(7,9)
>>> point2  = arcpy.Point(11,13)
>>> pList = [point,point1]
>>> array = arcpy.Array()
>>> array.extend(pList)
>>> array.replace(1,point2)
>>> point3  = arcpy.Point(17,15)
>>> array.insert(2,point3)
```

The `Array` object, when loaded with `Point` objects, can be used to generate other geometry objects such as `Polygon` and `Polyline` geometries. More information is available here: `http://desktop.arcgis.com/en/arcmap/latest/analyze/arcpy-classes/array.htm`

ArcPy Polyline objects

The `Polyline` object is generated with an `Array` object that has at least two `Point` objects. As in the IDLE example that follows, once an `Array` object has been generated and loaded with the `Point` objects, it can then be passed as a parameter to a `Polyline` object:

```
>>> point   = arcpy.Point(4,5)
>>> point1  = arcpy.Point(7,9)
>>> pList = [point,point1]
>>> array = arcpy.Array()
>>> array.extend(pList)
>>> pLine = arcpy.Polyline(array)
```

Now that the `Polyline` object has been created, its methods can be accessed. This includes methods to reveal the constituent coordinate points within the polyline, and other relevant information.

```
>>> pLine.firstPoint
<Point (4.0, 5.0, #, #)>
>>> pLine.lastPoint
<Point (7.0, 9.0, #, #)>
pLine.getPart()
<Array [<Array [<Point (4.0, 5.0, #, #)>, <Point (7.0, 9.0, #, #)>]>]>
>>> pLine.trueCentroid
<Point (5.5, 7.0, #, #)>
>>> pLine.length
5.0
```

```
>>> pLine.pointCount
2
```

This example `Polyline` object has not been assigned a spatial reference system, so the length is unitless. When a geometry object does have a spatial reference system, the linear and areal units will be returned in the linear unit of the system.

The `Polyline` object is also our first geometry object with which we can invoke geometry class methods that perform geospatial operations, such as buffers, distance analysis, and clips:

```
>>> bufferOfLine = pLine.buffer(10)
>>> bufferOfLine.area
413.93744395
>>> bufferOfLine.contains(pLine)
True
>>> newPoint = arcpy.Point(25,19)
>>> pLine.distanceTo(newPoint)
20.591260281974
```

Another useful method of `Polyline` objects is the `positionAlongLine` method. It is used to return a `PointGeometry` object, discussed later, at a specific position along the line. This position along the line can either be a numeric distance from the first point, or a percentage (expressed as a `float` from 0-1) when using the optional second parameter:

```
>>> nPoint = pLine.positionAlongLine(3)
>>> nPoint.firstPoint.X, nPoint.firstPoint.Y
(5.8, 7.4)
>>> pPoint = pLine.positionAlongLine(.5,True)
>>> pPoint.firstPoint.X,pPoint.firstPoint.Y
(5.5, 7.0)
```

There are a number of other methods available to Polyline objects. More information is available here:

```
http://desktop.arcgis.com/en/arcmap/latest/analyze/arcpy-class
es/polyline.htm
```

ArcPy Polygon objects

To create a `Polygon` object, an `Array` object must be loaded with `Point` objects and then passed as a parameter to the `Polygon` object. Once the `Polygon` object has been generated, the methods available to it are very useful for performing geospatial operations. The geometry objects can also be saved to disk using the ArcToolbox `CopyFeatures` tool. This IDLE example demonstrates how to generate a shapefile by passing a `Polygon` object and a raw string filename to the tool:

```
>>> import arcpy
>>> point1 = arcpy.Point(12,16)
>>> point2 = arcpy.Point(14, 18)
>>> point3 = arcpy.Point(11, 20)
>>> array = arcpy.Array()
>>> points = [point1,point2,point3]
>>> array.extend(points)
>>> polygon = arcpy.Polygon(array)
>>> arcpy.CopyFeatures_management(polygon, r'C:\Projects\Polygon.shp')
<Result 'C:\\Projects\\Polygon.shp'>
```

Polygon object buffers

`Polygon` objects, like `Polyline` objects, have methods to perform geospatial operations, such as buffers. For example, by passing a distance parameter to a `Polygon` object's `buffer` method, a new `Polygon` object representing the area of the buffered polygon will be generated in memory. The unit of the distance is determined by the spatial reference system. Internal buffers can be generated by supplying negative buffer distances; the buffer generated is the area within the `Polygon` object at the specified distance from the polygon perimeter.

Clips, unions, symmetrical differences, and more operations are available as methods, as are within or contains operations; even projections can be performed using the `Polygon` object methods as long as they have a spatial reference system object passed as a parameter. The following script creates two shapefiles with two separate spatial reference systems, each identified by a numeric code (**2227** and **4326** respectively) from the **EPSG** coding system:

```
import arcpyPoint  = arcpy.Point(6004548.231,2099946.033)
point1  = arcpy.Point(6008673.935,2105522.068)
point2  = arcpy.Point(6003351.355,2100424.783)Array = arcpy.Array()
array.add(point)
array.add(point1)
array.add(point2)
```

```
polygon = arcpy.Polygon(array, 2227)
buffPoly = polygon.buffer(50)
features = [polygon,buffPoly]
arcpy.CopyFeatures_management(features, r'C:\Projects\Polygons.shp')
spatialRef = arcpy.SpatialReference(4326)
polygon4326 = polygon.projectAs(spatialRef)
arcpy.CopyFeatures_management(polygon4326,
  r'C:\Projects\Polygon4326.shp')
```

There are a number of other methods available to `Polyline` objects. More information is available here:

http://desktop.arcgis.com/en/arcmap/latest/analyze/arcpy-class es/polygon.htm

The following screenshot shows how the second shapefile looks in the ArcCatalog **Preview** window:

Other Polygon object methods

Unlike the clip tool in the ArcToolbox, which can clip a geometry object to the shape of a polygon, the clip method of geometry objects requires an `Extent` object (another ArcPy class), and is limited to a rectangular envelope around the area to be clipped. To remove areas from a polygon, the difference method can work like the **Clip** or **Erase** tools in the ArcToolbox:

```
buffPoly = polygon.buffer(500)
donutHole =buffPoly.difference(polygon)
features = [polygon, donutHole]
arcpy.CopyFeatures_management(features, r"C:\Projects\Polygons2.shp")
```

The donut-hole-like result of the buffer and difference operation is copied to a shapefile using `arcpy.CopyFeatures_management`. Here, in a map document, the shapefile shows a donut hole carved from the buffer surrounding the original `Polygon` object:

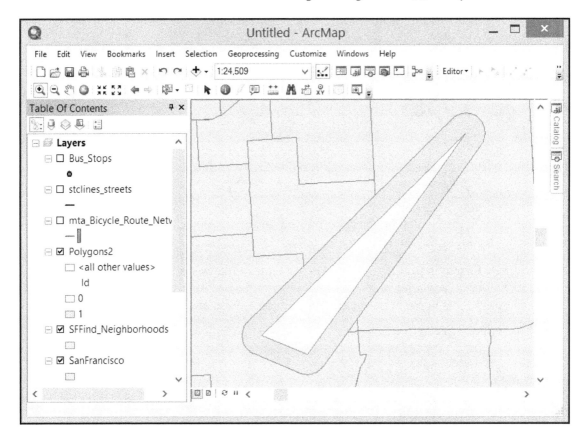

The AsShape method

Using a special method known as `AsShape`, a `Geometry` object can be created from **GeoJSON** or Esri JSON, two similar **JavaScript Object Notation (JSON)** spatial data formats, which have been adopted by the **Open Geospatial Consortium (OGC)** and Esri respectively.

GeoJSON is often used in web mapping and in the transfer of data from REST API endpoints. ArcGIS for server transfers data in the Esri JSON format when a request is made to a service URL. Databases and web map servers use GeoJSON to respond to requests.

As GeoJSON and Esri JSON are written in plain text, and use a special "key-value" format (which is based on Python dictionaries) to encode the data types geometry and attributes, the format makes it easy for both humans and computers to read and process the data it contains. The format is modeled on the Python dictionary data type using strings for the keys and lists for the values, and is easily read using built-in and third-party Python modules.

The ArcPy `AsShape` method makes it easy to incorporate spatial data encoded in these JSON formats into an ArcPy script. Let's explore how to convert GeoJSON and Esri JSON into a `Geometry` object.

First, the GeoJSON is passed to a variable as follows:

```
gjson = { "type" : "Point", "coordinates" : [-122.5, 37.8] }
```

Then, the variable is passed as a parameter to the ArcPy `AsShape` method like this:

```
geometry_point = arcpy.AsShape(gjson)
```

Once it is loaded, the GeoJSON object becomes a `Geometry` object with all of the methods available to the `Geometry` object.

The Esri JSON format is similar, but defines the required keys slightly differently. To load an Esri JSON object, a second parameter must be invoked to indicate that it is not a GeoJSON object. This Boolean parameter defaults to `False`, and must be set to `True` as shown next:

```
ejson = {"x": -122.5, "y": 37.8, "spatialReference": {
        "wkid": 4326}}
geometry_point = arcpy.AsShape(ejson, True)
```

More information about the `AsShape` method is available here:

http://desktop.arcgis.com/en/arcmap/latest/analyze/arcpy-funct
ions/asshape.htm

Generic geometry object

The generic geometry object, `arcpy.Geometry`, is useful for creating in memory a copy of the geometry of a feature class without first needing to know which type of geometry the feature class contains. Like all of the ArcPy geometry objects, its read methods include extraction of the data in many formats such as **JSON**, **WKT**, and **WKB**. Geometry properties such as area (if it is a polygon), centroid, extent, and the constituent points of each geometry are available.

Here is an example of reading the geometry of a feature class into memory using the `CopyFeatures` tool:

```
import arcpy
cen2010 = r'C:\Projects\SanFrancisco.gdb\SanFrancisco\CensusBlocks2010'
blockPolys = arcpy.CopyFeatures_management(cen2010, arcpy.Geometry())
```

The variable `blockPolys` is a Python list containing all of the geometry's data loaded into it; in this case, it is census blocks. The list of blocks can be analyzed and processed one by one using iteration.

ArcPy PointGeometry objects

The `PointGeometry` object is very useful for performing these same geospatial operations with points, which are not available with the Point objects. When a cursor is used to retrieve shape data from a feature class with a `PointGeometry` type, the shape data is returned as a `PointGeometry` object. While `Point` objects are required to construct all other geometry objects when a cursor is not used to retrieve data from a feature class, it's the `PointGeometry` object that is used to perform point geospatial operations.

```
>>> import arcpy
>>> arcpy.Point(1.0, 2.0)
<Point (1.0, 2.0, #, #)>
>>> point = arcpy.Point(1.0, 2.0)
>>> point.buffer
Traceback (most recent call last)
 File "<pyshell#3>", line 1, in <module>
   point.buffer
```

```
AttributeError: 'Point' object has no attribute 'buffer'
>>> point_geometry = arcpy.PointGeometry(point)
>>> point_geometry.buffer
<bound method PointGeometry.buffer of <PointGeometry object at
0xce903f0[0xce6fe60]>>
>>> point_geometry.buffer(10)
<Polygon object at 0xce6ff10[0xce6fe40]>
```

Rewriting the bus stop analysis

Using geometry objects and their methods, we can rewrite the bus stop analysis. The new script, `Chapter4_1.py`, will iterate through the bus stops, and find all the census blocks intersecting a buffer around each stop. The `PointGeometry` objects are returned from the shape field of a point feature class by a data access `SearchCursor`. Using the returned `PointGeometry` objects, each bus stop is buffered to create a polygon geometry. In the new bus stop analysis, this will replace the need to use the ArcToolbox **Buffer** tool to create the 400-foot buffers around each stop. The buffer polygon is then analyzed for overlap with the census block geometries. Any blocks found to be overlapping with the buffer polygon are intersected with the buffer to produce a new dataset with a polygon geometry and census block attributes. The results are processed to produce an average population served by each bus stop, and written to a CSV file.

The new script starts with importing libraries and defining the variables. It is important to ensure that the file paths of the data sources match your own:

```
# Import libraries and define variables
import arcpy, csv
busStops = r"C:\Projects\SanFrancisco.gdb\SanFrancisco\Bus_Stops"
census2010 =
    r"C:\Projects\SanFrancisco.gdb\SanFrancisco\CensusBlocks2010"
csvname = r"C:\Projects\StationPopulations.csv"
sql = "NAME = '71 IB' AND BUS_SIGNAG = 'Ferry Plaza'"
headers = 'Bus Line Name','Bus Stop ID', 'Average Population'
```

The rewritten script given next uses a dictionary to collect the buffer objects, and then searches the census blocks using another Search Cursor. To access the shape field using the `SearchCursor`, the `SHAPE@` token is passed as one of the fields:

```
dataDic = {}
with arcpy.da.SearchCursor(busStops,'NAME','STOPID','SHAPE@'], sql) as
    cursor:
        for row in cursor:
            linename = row[0]
            stopid = row[1]
```

```
    shape = row[2]
    dataDic[stopid] = shape.buffer(400), linename
```

Now that the data has been retrieved and the buffers have been generated using the buffer method of the `PointGeometry` objects, the buffers can be compared against the census block geometry using iteration and a `SearchCursor`. There will be two geospatial methods used in this analysis: `overlap` and `intersect`.

The `overlap` method is a Boolean operation, which returns a value of true or false when one geometry is compared against another. The `intersect` method is used to get the actual area of intersection along with finding the population of each block. Using the `intersect` method requires two parameters: a second geometry object, and an integer indicating which type of geometry to return (**1** for point, **2** for line, and **4** for polygon). We want the polygonal area of intersect returned to have an area of intersection available along with the population data:

```
# Intersect census blocks and bus stop buffers
processedDataDic = {}
for stopid in dataDic.keys():
    values = dataDic[stopid]
    busStopBuffer = values[0]
    linename = values[1]
    blocksIntersected = []
    with arcpy.da.SearchCursor(census2010,
      ['BLOCKID10','POP10','SHAPE@']) as cursor:
        for row in cursor:
            block = row[2]
            population = row[1]
            blockid = row[0]
            if busStopBuffer.overlaps(block) ==True:
                interPoly = busStopBuffer.intersect(block,4)
                data = row[0],row[1],interPoly, block
                blocksIntersected.append(data)
    processedDataDic[stopid] = values, blocksIntersected
```

This preceding portion of the script iterates through the blocks, and intersects against the buffered bus stops. Now that we can identify the blocks that touch the buffer around each stop, and the data of interest has been collected into the dictionary, it can be processed, and the average population of all the blocks touched by the buffer can be calculated as follows:

```
# Create an average population for each bus stop
dataList = []
for stopid in processedDataDic.keys():
    allValues = processedDataDic[stopid]
    popValues = []
    blocksIntersected = allValues[1]
```

```
    for blocks in blocksIntersected:
        popValues.append(blocks[1])
    averagePop = sum(popValues)/len(popValues)
    busStopLine = allValues[0][1]
    busStopID = stopid
    finalData = busStopLine, busStopID, averagePop
    dataList.append(finalData)
```

Now that the bus stop average population data has been created and added to a list, it can be outputted to a spreadsheet using the createCSV module that we created in Chapter 3, *ArcPy Cursors - Search, Insert, and Update*:

```
# Generate a spreadsheet with the analysis results
def createCSV(data, csvname, mode ='ab'):
    with open(csvname, mode) as csvfile:
        csvwriter = csv.writer(csvfile, delimiter=',')
        csvwriter.writerow(data)

createCSV(headers, csvname, 'wb')
for data in dataList:
    createCSV(data, csvname)
```

The data has been processed and written to the spreadsheet.

Adding to the analysis

There is one more step that we can take with the data, and that is to use the area of intersection to create a proportional population value for each buffer. Let's redo the processing of the data to include the proportional areas:

```
dataList = []
for stopid in processedDataDic.keys():
    allValues = processedDataDic[stopid]
    popValues = []
    blocksIntersected = allValues[1]
    for blocks in blocksIntersected:
        pop = blocks[1]
        totalArea = blocks[-1].area        # using the area property
        interArea = blocks[-2].area
        finalPop = pop * (interArea/totalArea)
        popValues.append(finalPop)
    averagePop = round(sum(popValues)/len(popValues),2)
    busStopLine = allValues[0][1]
    busStopID = stopid
    finalData = busStopLine, busStopID, averagePop
    dataList.append(finalData)
```

This change takes advantage of the polygon geometry objects and their area property. Using this property, the total area of the census block and the area of intersection with the buffer polygon around the bus stop can be accessed, and used in an equation that estimates the percentage of the census block population that is within the service area of the bus stop.

The rewritten bus stop analysis script now takes full advantage of ArcPy geometry objects and data access cursors. The analysis runs in memory, avoids producing any intermediate datasets, and creates custom tabular output.

Summary

In this chapter, we discussed in detail the use of ArcPy geometry objects. These varied objects have similar methods, and are, in fact, sub-classed from the same Python class. They are useful for performing in-memory geospatial analysis, which avoids having to read and write data from the hard drive, and also skips creating any intermediate data.

ArcPy geometry objects are an important part of automating geospatial workflows. Combining them with Search Cursors makes ArcPy more useful than any earlier implementation of Python scripting tools for ArcGIS. In the next chapter, we will convert the raw script into a script tool that can be executed directly from the ArcToolbox or a personal toolbox in a geodatabase.

5

Creating a Script Tool

Now that the basics of creating and executing ArcPy scripts have been covered, we need to take the next step, and create reuseable "script tools". Creating script tools will allow for greater code reuse, and will make it easy to create custom tools for other GIS analysts and customers. With a Python script "back end", or code, and a familiar ArcGIS tool "front end" or graphical user interface (GUI), the custom script becomes a reliable tool for all users within a GIS work shop.

This chapter will cover the following topics:

- Adding parameters to scripts to accept input, and produce output as required by the user
- Creating a custom tool front end and a custom toolbox
- Setting the parameters of the tool front end to allow it to pass arguments to the code back end

Adding dynamic parameters to a script

The scripts we have generated in previous chapters have all had "hard-coded" inputs. The input values were written in the script as strings or numbers, and assigned to variables. While they can be updated manually to replace the input and output file paths and SQL statements, programmers should aim to create reusable code. Scripts should be designed to be dynamic, accepting file paths and other inputs as parameters or arguments. Like Python functions, scripts should accept parameters, process the data, and produce a result.

Python was designed with this in mind, and dynamic parameters can be passed to scripts when executed. How are parameters, also known as arguments, passed to the script? There are a few methods. When running a script in the command line, script parameters are passed to the script separated by spaces, in the order expected by the script.

In this example, python is called (if `python.exe` is in the Path environment variable; see `Chapter 1`, *Introduction to Python for ArcGIS*) and the script `testcmd.py` is passed to it, and then the parameters as passed to the script, in order and separated by spaces:

```
#testcmd.py
import sys
scriptpath = sys.argv[0]
parameter1 = sys.argv[1]
parameter2 = sys.argv[2]
print scriptpath, parameter1, parameter2
print type(scriptpath), type(parameter1), type(parameter2)
```

The script must be within the current folder, or it must be passed as a complete file path. Parameters can be strings, integers, and floats and cannot contain spaces, unless the parameter is in quotes. The parameters passed will be converted to strings and must be converted back to the correct data type within the script. If we pass feature class paths to a script, they would be strings, and could be used without conversion.

With script tools, we avoid having to use a command line interface and instead create a GUI with entry fields that have formatting, logic and parameters built-in. This avoids the need to use a command line to run dynamic ArcPy scripts as well as data conversion issues.

Accessing the passed parameters

Within scripts, the `sys` module represents the Python executable itself. The method `sys.argv` is used to access the arguments passed to the script, as shown earlier. While the designers of ArcPy and its predecessor module, `arcgisscripting` initially took advantage of `sys.argv` to build dynamic scripts, a built-in ArcPy method for accessing script parameters has been created: `arcpy.GetParameterAsText`.

Either method can be used when writing Arcpy scripts, and as both are found in example scripts on the web, it is important to recognize the `sys.argv` method and `arcpy.GetParameterAsText`. The difference between the two methods is that `sys.argv` provides the dynamic arguments as a list. Members of the list are accessed using list indexing, and assigned to variables. The `arcpy.GetParameterAsText` is a function which accepts an index number parameter and returns that parameter. The index number of each parameter reflects the order of the parameter passed to the script: the first parameter is 0, the next is 1, and so on.

Here is an example of `arcpy.GetParameterAsText`. The first parameter passed to the script is accessed with index 0, while the second is accessed with index 1:

```
import arcpy
busStops = arcpy.GetParameterAsText(0)
censusBlocks2010 = arcpy.GetParameterAsText(1)
```

When using `sys.argv` the indexing for parameters starts at index 1. This is because the file path of the script itself is at `sys.argv[0]` for any script.

```
import arcpy, sys
busStops = sys.argv[1]
censusBlocks2010 = sys.argv[2]
```

If the order of the parameters is adjusted in the tool front end, this adjustment must be reflected in the code as well. Make sure that the index values of `arcpy.GetParameterAsText` in the script match the order of the parameters passed to ensure correct code execution.

Either one of these methods used would work with command line arguments passed to a script in this manner:

Displaying script messages using arcpy.AddMessage

It is important to receive feedback from scripts to assess the progress of the script, as it performs an analysis. As basic as this would seem, Python scripts and programming languages, in general, do not, by default, provide any feedback except for errors and the termination of the script. This can be a bit discouraging to the novice programmer, as all built-in feedback is negative.

To alleviate this lack of feedback, the use of print statements allow the script to give reports on the progress of the analysis as it runs. However, when using a script tool, print statements do not have any effect. They will not be displayed anywhere, and are ignored by the Python executable. To display messages in the script console when script tools are executed, ArcPy has a method: `arcpy.AddMessage`.

The `arcpy.AddMessage` statements are added to scripts wherever feedback is required by the programmer. The feedback required is passed to the method as a parameter and displayed, whether it be a list, string, float, or integer. The `arcpy.AddMessage` makes it easy to check on the results of analysis calculations to ensure that the correct inputs are used, and that the correct outputs are produced. As this feedback from the script can be a powerful debugging tool, use `arcpy.AddMessage` whenever there is a need for feedback from the script tool.

```
arcpy.AddMessage('This is a message')
```

 Statements passed to `arcpy.AddMessage` must be strings and will only display when the script is run as a script tool.

Adding dynamic components to the script

To start making the script into a script tool, we should first copy the script that we created in Chapter 4, *ArcPy Geometry Objects and Cursors*, `Chapter4_1.py` as `Chapter5_1.py` in a new folder called `Chapter5`. We can then start replacing the hard-coded variables with dynamic variables using `arcpy.GetParameterAsText`. There is another ArcPy method called `GetParameter`, which accepts the inputs as an object, but, for our purpose, we need to use `GetParameterAsText`.

By adding `arcpy.GetParameterAsText` and `arcpy.AddMessage` to the script, we will take the first step towards creating a script tool. Care must be taken to ensure that the variables created from the dynamic parameters are in the correct order, as reordering them can be time consuming. Once the parameters are added to the script, and hard-coded portions of the script are replaced with variables, the script is ready to become a script tool.

First, move all of the variables that are hard-coded to the top of the script. Then, replace all of the assigned values with `arcpy.GetParameterAsText`, making them dynamic values. Each parameter is referred to using zero-based indexing; however, they are passed to a function individually instead of as a member of a list, as shown next:

```
#Chapter 5_ScriptTool.py
import arcpy, csv
busStops = arcpy.GetParameterAsText(0)
censusBlocks2010 = arcpy.GetParameterAsText(1)
censusBlockField = arcpy.GetParameterAsText(2)
csvname = arcpy.GetParameterAsText(3)
headers = arcpy.GetParameterAsText(4).split(',')
sql = arcpy.GetParameterAsText(5)
keyfields = arcpy.GetParameterAsText(6).split(';')

censusFields = ['BLOCKID10',censusBlockField, 'SHAPE@']
if "SHAPE@" not in keyfields:
    keyfields.append("SHAPE@")
```

As you can see from the variable `keyfields` and the variable `headers` in the preceding code, some further processing must be applied to certain variables, as not all of them are going to be used as strings. In this case, a list is created from the string passed to the variable `keyfields` by using the string functions `split`, and splitting the string on every semicolon, while the headers variable is created by splitting on the commas. To other variables, such as the `censusBlockField` variable and the variable `keyfields`, the SHAPE@ keyword is added, because it will be required each time the analysis is run. If a particular field is required for each run of the data, such as the `BLOCKID10` field, it can remain hard-coded in the script, or optionally, could become its own field parameter in the script tool.

With the variables in place, the script tool is ready to become part of a custom toolbox in a geodatabase or in ArcToolbox. Let's create the front end GUI for the script tool to collect and pass parameters to the script.

Creating a script tool

Script tools are common within GIS. ArcToolbox includes many script tools, and ArcGIS for Desktop allows for personal toolboxes with a mix of models and script tools. This example will demonstration how to create the front end graphic user interface or GUI that will pass the correct parameters in the correct order to the script.

The first step is to create a custom tool box to hold the script tool. Perform the steps listed here:

1. Open up ArcCatalog, and right-click on the `SanFrancisco.gdb` file Geodatabase.
2. Select New and then Toolbox from the menu.
3. Call the toolbox Chapter5Tools.
4. Right-click on Chapter5Tools, and select Add, then select Script.

The following menu will appear, allowing you to set up the script tool. Add a title **Name** with no spaces, and a label as well as a description. I prefer to run script tools in the foreground to see the messages it passes, but it is not necessary, and can be annoying when needing to start a script and still work on other tasks. Click **Next** once the menu has been filled out:

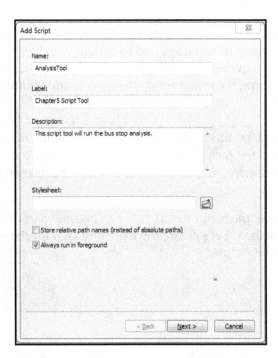

Locate the script that will run once the script tool is complete. Use the file menu to find the script:

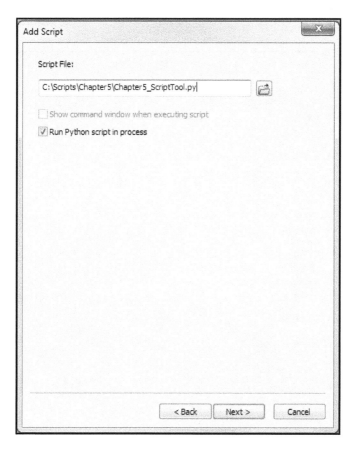

The next menu contains an entry field and a file dialog button, allowing the user to find the script to which the parameters collected will be passed. Use the file dialog to navigate to and select the script, and make sure that **Run Python script in process** is checked. Click **Next** once the script has been identified.

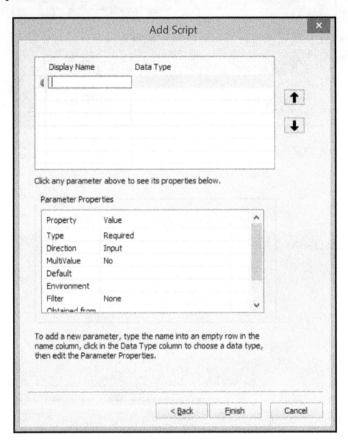

Labeling and defining parameters

The next dialog box is the most important one. It is where the parameters to be passed are labeled and their data types are defined. Care must be taken to choose the correct data type for each parameter, as there are multiple data types that can be supplied for some of the parameters. Also, properties for each parameter will be defined; this information will characterize the data to be collected, and help to make it possible for the data to be in the correct format as well as the correct data type.

Start by adding the **Display Name** for each parameter to be collected. The **Display Name** should explain the type of input that is required. For instance, the first parameter will be the Bus Stop feature class, so the display name could be Bus Stop feature class.

 Make sure to add the display names in the order that they will be passed to variables in the script. Use the arrows on the right to reorder them when required.

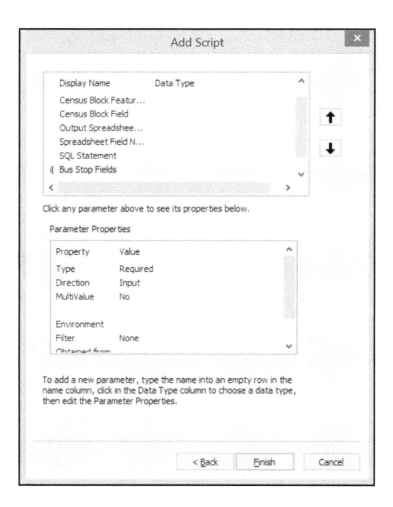

Adding data types

Next, add in the **Data Type** for each parameter. This is intricate, because there will be a large list of data types to choose from, and often, there are a few types that would work for many parameters. For instance, if a shapefile parameter is created, it would allow the user to select a shapefile as expected. However, it might be better to use the feature class data type, as then, both shapefiles and feature classes could be used in the analysis. Here is an overview of the parameters:

Parameter	Data Type	Derived from?	Default
Bus Stop Feature Class	Feature Class		
Census Blocks Feature Class	Feature Class		
Census Block Field	Field	Census Block	
Output Spreadsheet	String		`BusStopAnalysis.csv`
Spreadsheet Field Headers	String		Bus Line Name, Bus Stop ID, Population
SQL Statement	SQL Expression	Bus Stop	
Bus Stop Fields	Field	Bus Stop	

Adding the Bus Stop feature class

The first parameter is the Bus Stop feature class, and it should be a feature class data type. Click on the **Data Type** field next to the display name, and a drop-down list will appear. Once the data type is selected, check out the parameter properties after the list of parameters. For the Bus Stop feature class, the defaults will be acceptable, as the feature class is required, is not a multi-value, has no default or environment settings, and is not obtained from any other parameter:

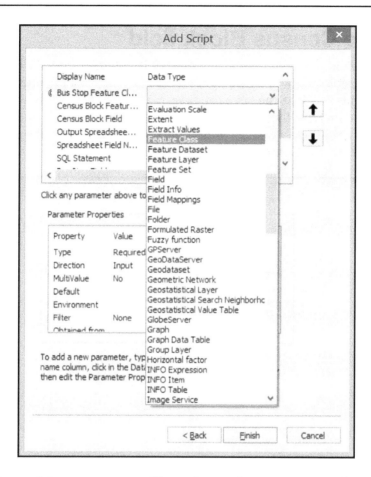

 Some of the parameters will require another parameter to be selected first, as their selections are values obtained from the first parameter. The Census Block Field parameter are obtained from the Census Block feature class, while the SQL statement parameter and the Bus Stop Field parameter are obtained from the Bus Stop feature class.

Adding the Census Block feature class

The Census Block feature class is similar to the Bus Stop feature class. It will be a **Feature Class** data type, which allows both shapefiles and feature classes to be selected, and there is no need to adjust the default parameter properties. Once the data type has been set, the Census Block parameter is ready for use.

Adding the Census Block field

The Census Block field parameter has a new twist: it is obtained from the Census Block feature class parameter, and will only be populated once that first parameter has been created. To make this possible, the Obtained from parameter property will have to be set. Select **Field** as the data type, then click on the blank area next to the Obtained from parameter property, and select **Census_Block_Feature_Class**. This will create a list of the fields contained within the Census Block feature class:

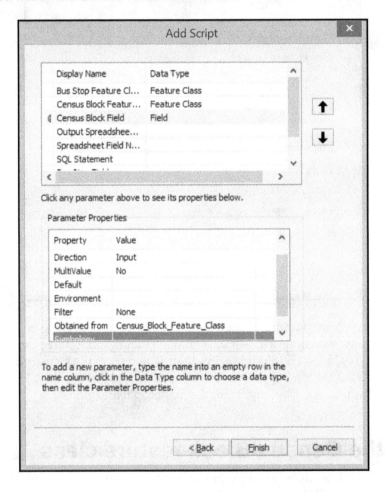

Adding the output spreadsheet

As the spreadsheet that will be produced from the analysis run by the script tool is a **comma-separated value** (**CSV**) file, select **String** as the **Data Type**, as we will just use the string passed as the file path. Setting the **Default** parameter property to a file name can save time. The string will be passed to the `createCSV` function to produce the output spreadsheet:

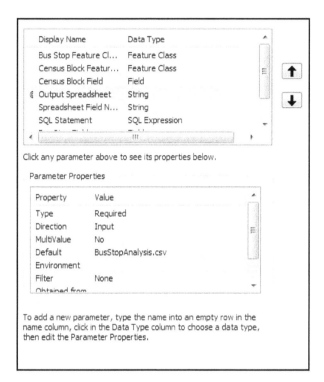

Adding the spreadsheet field names

The field names chosen as headers for the output spreadsheet should be String data type, with the field names separated by commas and no spaces. The script uses the `String` data type's `split` method to separate the field names. Passing a comma to the split method separates the parameter by "splitting" the input string on the commas to create a list of field names. The list of field names will be used as a header by the `csv` module when creating the spreadsheet.

Adding the SQL Statement

The **SQL Statement** parameter will require the helpful SQL Expression Builder menu, and should, therefore, be an SQL Expression data type. The SQL Expression Builder will use a field obtained from the Bus Stop feature class. Set the **Obtained from** parameter property to the Bus Stop feature class by clicking on that property, and selecting **Bus_Stop_Feature_Class** from the drop-down list.

Adding the bus stop fields

The final parameter is the **Bus Stop Fields** parameter, which is a **Field** data type that will be obtained from the Bus Stop feature class. Change the **MultiValue** parameter property from **No** to **Yes** to allow the user to select multiple fields. Also remember to set the **Obtained from** parameter property to **Bus_Stop_Feature_Class** so that the fields are populated from the Bus Stop feature class parameter as shown:

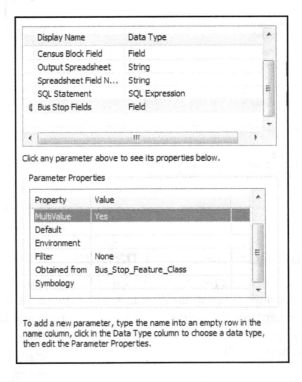

Now that all the parameters have been described and their properties have been set, the script tool is ready. Click on **Finish** to close the menu.

Inspecting the final script

Once all of the parameters have been collected, the variables are then used to replace the hard-coded field lists or other static script elements with the new dynamic parameters collected from the script tool. In this manner the script becomes a valuable tool, which can be used for multiple data analysis, as the fields to be analyzed are now dynamic.

```python
# Import libraries and define variables to be derived from parameters
import arcpy, csv
busStops = arcpy.GetParameterAsText(0)
censusBlocks2010 = arcpy.GetParameterAsText(1)
censusBlockField = arcpy.GetParameterAsText(2)
csvname = arcpy.GetParameterAsText(3)
headers = arcpy.GetParameterAsText(4).split(',')
sql = arcpy.GetParameterAsText(5)
keyfields = arcpy.GetParameterAsText(6).split(';')
dataDic = {}
censusFields = [ 'BLOCKID10',censusBlockField,'SHAPE@']
if "SHAPE@" not in keyfields:
    keyfields.append("SHAPE@")

# Add message is used instead of print
arcpy.AddMessage(busStops)
arcpy.AddMessage(censusBlocks2010)
arcpy.AddMessage(censusBlockField)
arcpy.AddMessage(csvname)
arcpy.AddMessage(sql)
arcpy.AddMessage(keyfields)

x = 0
with arcpy.da.SearchCursor(busStops, keyfields, sql) as cursor:
    for row in cursor:
        stopid = x
        shape = row[-1]
        dataDic[stopid] = []
        dataDic[stopid].append(shape.buffer(400))
        dataDic[stopid].extend(row[:-1])
        x+=1
processedDataDic = {}
for stopid in dataDic.keys():
    values = dataDic[stopid]
    busStopBuffer = values[0]
    blocksIntersected = []
    with arcpy.da.SearchCursor(censusBlocks2010, censusFields) as
      cursor:
        for row in cursor:
            block = row[-1]
```

```
                population = row[1]
                blockid = row[0]

                if busStopBuffer.overlaps(block) ==True:
                        interPoly = busStopBuffer.intersect(block,4)
                        data = population,interPoly, block
                        blocksIntersected.append(data)
        processedDataDic[stopid] = values, blocksIntersected
dataList = []
for stopid in processedDataDic.keys():
    allValues = processedDataDic[stopid]
    popValues = []
    blocksIntersected = allValues[-1]
    for blocks in blocksIntersected:
        pop = blocks[0]
        totalArea = blocks[-1].area
        interArea = blocks[-2].area
        finalPop = pop * (interArea/totalArea)
        popValues.append(finalPop)
    averagePop = round(sum(popValues)/len(popValues),2)
    busStopLine = allValues[0][1]
    busStopID = stopid
    finalData = busStopLine, busStopID, averagePop
    dataList.append(finalData)
def createCSV(data, csvname, mode ='ab'):
    with open(csvname, mode) as csvfile:
        csvwriter = csv.writer(csvfile, delimiter=',')
        csvwriter.writerow(data)

createCSV(headers, csvname, 'wb')
for data in dataList:
    createCSV(data, csvname)
```

The variable **x** was introduced to eliminate the dependency of `dataDic` on the `STOPID` field as it had in `Chapter 4`, *ArcPy Geometry Objects and Cursors*. Instead, the key are just numbers.

Try making the buffer distance dynamic by adding it as a variable in the script and as an input on the front end!

Running the script tool

Now that the front end has been designed to accept parameters from a user, and the back end script is ready to accept the parameters from the front end, the tool is ready to be executed. Double-click on the script tool in the toolbox to open the tool dialog box, and fill it out:

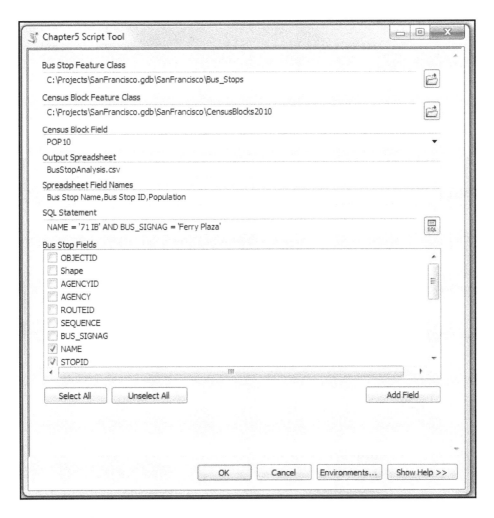

One optional adjustment would be to add an `arcpy.AddMessage` line where the average population is calculated. By doing this, the individual block population would be displayed, and the script console would give feedback about the progress of the script.

Insert the following in the script just after the line where the variable `finalData` is defined:

```
finalData = busStopLine, busStopID, averagePop
arcpy.AddMessage(finalData)
```

The feedback provided by this line will make it obvious that the script is working, which is useful when the script executes a long analysis. When performing a long analysis, it is good practice to provide feedback to the user so that they can see that the script is working as expected. Pass newline characters *n* as parameters to `arcpy.AddMessage` when there is a large amount of data being passed to `arcpy.AddMessage`. This will break up the data into discrete chunks, and make it easier to read.

Summary

In this chapter, you learned how to convert a script into a permanent and shareable script tool, which can be used by an ArcGIS user with no programming experience. By creating a front end using the familiar ArcGIS tool interface, and then passing parameters to custom built tools that employ ArcPy, GIS programmers can combine the ease of the ArcToolbox and the power of Python.

In the next chapter, we will explore using ArcPy to control the export of maps from map documents. By adjusting map elements such as titles and legends, we can create dynamic map outputs to display the data produced by map analysis.

6
The arcpy.mapping Module

Creating maps is an art, one that can be learned through years of dedicated study of cartography. The visual display of information is both exciting and difficult, and can be a rewarding part of the daily work flow of geospatial professionals. Once the basics have been learned and then mastered, cartographic output becomes a constant battle to produce more maps at a faster pace. ArcPy once again has a powerful solution: the `arcpy.mapping` module.

By allowing for the automatic manipulation of all map components, including the map window, the layers, the legend, and all text elements, `arcpy.mapping` makes creating, modifying, and outputting multiple maps fast and simple. Map book creation, another important skill for geospatial professionals, is also made easy using the module. In this chapter, we will discuss some basic functionality available through `arcpy.mapping` and use it to output a map book of bus stops and their surrounding census blocks.

This chapter will cover the following topics:

- Inspecting and updating Map Document (MXD) layer data sources
- Exporting MXDs to PDF or other image formats
- Adjusting map document elements
- The `arcpy.mapping` layer objects
- The `arcpy.mapping` data frame objects
- Creating dynamically scaled maps

Using ArcPy with map documents

Esri designed the `arcpy` module to not only work with data but also included the `arcpy.mapping` module for direct interaction with MXDs and the layers they contain. This module opens up a multitude of map automation possibilities. A script might aid in identifying broken layer links, update the data source of these layers, and apply new color schemes to layers. Another script might use a map template to create a set of maps, one from each feature class in a feature dataset. A third script could create a map book from an MXD, moving from cell to cell in a grid layer to output the pages of the book, or even calculating the coordinates on the fly. Dynamically created maps, based on data from a fresh analysis, can be created as data is produced. The `arcpy.mapping` module moves the `arcpy` module from helpful to instrumental in any geospatial work flow.

To investigate the utility of the `arcpy.mapping` module, we'll need an MXD template document. I've included with the code samples a folder containing the data and MXDs that we will use for the exercises in this chapter. It includes the data from our San Francisco bus stops analysis, which we will continue and extend to include maps.

Interacting with map document elements

Map document elements include layers, legends, text, data frames, all of which are controllable using `arcpy.mapping`. All of these elements are Python objects, with methods and properties that mirror those available through the various properties menus.

To access these elements, use the `arcpy.mapping.MapDocument` method. It connects to the map document by accepting an MXD file path as a string:

```
import arcpy
mxdPath = r'C:\Projects\MXDs\BrokenLinks.mxd'
mxdObject = arcpy.mapping.MapDocument(mxdPath)
```

Data frames

Use the `ListDataFrames` method to access `DataFrame` objects as shown in the following code:

```
dataFrame = arcpy.mapping.ListDataFrames(mxdObject, "")[0]
```

With `DataFrame` objects comes control over data frames - the map "window"- and the ability to zoom and pan over an area of interest.

As map layers are organized within data frames in MXDs, the `DataFrame` objects are required for access to layer objects.

Pan and zoom methods

There are many following `DataFrame` object methods used to shift the data frame window to an area of interest:

- The first is the data frames's `zoomToSelectedFeatures` method:

```
arcpy.SelectLayerByAttribute_management(bus_stop_lyr,
  "NEW_SELECTION","NAME='71IB'")
dataFrame.zoomToSelectedFeatures()
```

- The second is to create an `Extent` geometry and assign the data frame extent to that geometry:

```
extentGeom = arcpy.Extent(100, 20, 400, 200)
dataFrame.extent = extentGeom.extent
```

- The third is to use the `panToExtent` method:

```
layersList = arcpy.mapping.ListLayers
(mxdObject,"",dataFrame)
dataFrame.panToExtent(layersList.extent)
```

- The fourth is to assign the data frame's `extent` property to the extent of another layer or an ArcPy geometry object:

```
layersList = arcpy.mapping.ListLayers(mxdObject,"",dataFrame)
censusBlocks = layersList[0]
dataFrame.extent = censusBlocks.extent
```

- The `scale` property is very useful as it can be set to a restricted set of scales, or as a multiplier of the current scale. When using the `scale` property, remember to use the `arcpy.RefreshActiveView` method, to enable the changes to the scale of the data frame window to take effect. In this case the multiplier is `1.1`, but it could be any value:

```
dataFrame.scale = dataFrame.scale * 1.1
arcpy.RefreshActiveView()
```

Using the arcpy.mapping module to control layer objects

Turning layer visibility on and off, adding new layers, adjusting layer order, or even styling are all accomplished using the layer object properties. However, many layer properties must still be adjusted in ArcMap.

When referencing an `arcpy.mapping.MapDocument` object, the layers within the map document can be called as a list, and accessed individually using zero-based indexing. The `arcpy.mapping` layer objects are used to control the properties of layers within map document data frames.

This code will print the list of layer objects contained within the data frame called layers in an MXD:

```
import arcpy
mxdPath = r'C:\Projects\MXDs\MapDocument1.mxd'
mxdObject = arcpy.mapping.MapDocument(mxdPath)
dataFrame = arcpy.mapping.ListDataFrames(mxdObject, "Layers")[0]
layersList = arcpy.mapping.ListLayers(mxdObject,"",dataFrame)
print layersList
```

The layers within the data frame called "Layers" are assigned to the variable `layersList` using the `ListLayers` method. Individual layers can be accessed using zero-based indexing. Once they have been accessed within the list and either assigned to a variable or placed inside a for loop, the properties, and methods of the layer objects can be utilized.

The second parameter of the `ListLayers` method is empty here, but does not have to be. It is a wild card parameter that will limit the returned Layer objects to those that match the pattern of the wildcard. For instance, `*Stops` would return all layers with the name "Stops" at the end. Asterisks can be used to find layers with the word at the beginning, middle, or end of the layer name.

Layer object methods and properties

Layer object properties and methods are either read only, meaning they can be checked but not adjusted using scripting, or are read and write, meaning they can be adjusted from the script. Let's explore a number of these properties and methods and see how they can be used to control the look and feel of the maps produced from the map document, as well as the data from the script analysis.

Data source

Sometimes it is necessary to confirm the data source of a layer. While data source is a read only property, it can be adjusted using the methods described in the preceding section:

```
layersList = arcpy.mapping.ListLayers(mxdObject,"",dataFrame)
layerStops = layersList[0]
layerBlocks = layersList[3]
print layerStops.dataSource, layerBlocks.dataSource
```

Name or description

The name and description fields are read and write, meaning that how a layer is named in the **Table of Contents** (and therefore in a legend) can be adjusted using these properties:

```
layersList = arcpy.mapping.ListLayers(mxdObject,"",dataFrame)
layerStops.description = "SF Bus Stop"
layerBlocks.name = "Census Block Data"
```

Visibility

Controlling a layer's visibility means turning the layer on or off within the MXD. This can be very useful for controlling the output of a map and deciding which layers should appear for each output page produced. To turn off the visibility of a layer, set the `visible` property to `False`. To turn it on, set the property to `True`:

```
layersList = arcpy.mapping.ListLayers(mxdObject,"",dataFrame)
layerStops = layersList[0]
layerBlocks = layersList[3]
layerStops.visible = True
layerBlocks.visible = False
```

Definition queries

An important property of layer objects is the ability to set a definition query. A definition query is an SQL statement that selects a portion or subset of the data within the layer's data source, so that only the selected part of the data is available for display or other data operations (buffers, intersections, and so on):

```
layersList = arcpy.mapping.ListLayers(mxdObject,"",dataFrame)
busStops = layersList[0]
busStops.definitionQuery = "NAME = '71 IB'"
```

Inspecting and replacing layer sources

An important `arcpy.mapping` use is identifying and fixing the broken links between layers in a map document and their data sources. Layer symbology and GIS data storage are separated, meaning that layer data sources are often moved. The `arcpy.mapping` module offers a quick, though imperfect, solution.

This solution depends on a number of methods included in the `arcpy.mapping` module. First we will need to identify the broken links, and then we will fix them. To identify the broken links we will use the `ListBrokenDataSources` method included in the `arcpy.mapping` module.

The ListBrokenDataSources method

The `ListBrokenDataSources` method requires an MXD path to be passed to the `MapDocument` method of `arcpy.mapping`. Once the map document object has been created, it is passed to the `ListBrokenDataSources` method, and a list will be generated containing layer objects, one for each layer with a broken link. The layer objects have a number of properties available to them. Using these properties, let's print out the name and data source of each layer using the `name` and `dataSource` properties of each object:

```
import arcpy
mxdPath = r'C:\Projects\MXDs\BrokenLinks.mxd'
mxdObject = arcpy.mapping.MapDocument(mxdPath)
brokenLinks = arcpy.mapping.ListBrokenDataSources(mxdObject)
for link in brokenLinks:
  print link.name, link.dataSource
```

Fixing the broken links

Now that we have identified the broken links, the next step is to fix them. The script must be improved to replace the data sources of each layer so that they point at the actual location of the data source.

Both layer objects and map document objects have built-in methods to fix broken links. If all of the data sources for an MXD have been moved, it is best to use the MXD object to repair the broken links. For example, if the data sources have all been moved into a new folder called `NewData`, we will employ the `findAndReplaceWorkspacePaths` method to repair the links:

```
oldPath = r'C:\Projects\OldData'
newPath = r'C:\Projects\NewData'
mxdObject.findAndReplaceWorkspacePaths(oldPath,newPath)
```

```
mxdObject.save()
```

As long as the data sources are still in the same format (that is, shapefiles are still shapefiles and feature classes are still feature classes), the find and replace workspace paths method will work.

If the data source types have been changed (that is shapefiles are imported into a File Geodatabase), the `replaceWorkspaces` method is used. The `replaceWorkspaces` method requires workspace type as a parameter:

```
oldPath = r'C:\Projects\OldSHPs'
oldType = 'SHAPEFILE_WORKSPACE'
newPath = r'C:\Projects\NewData\SanFrancisco.gdb'
newType = 'FILEGDB_WORKSPACE'
mxdObject.replaceWorkspaces(oldPath,oldType,newPath,newType)
mxdObject.save()
```

Fixing the links of individual layers

If individual layers do not share a data source, the layer objects will be adjusted using the `findAndReplaceWorkspacePath` method available to layers. This method is similar to the `replaceWorkspaces` method used in the preceding code snippet, but it will only replace the data source of the layer object it is applied to instead of all of the layers. When combined with a dictionary, the layer data sources can be updated using the layer name property:

```
import arcpy
layerDic = {'Bus_Stops':
  [r'C:\Projects\OldData',r'C:\Projects\NewData'],
    'stclines_streets':
    [r'C:\Projects\OldData',r'C:\Projects\NewData']}
mxdPath = r'C:\Projects\MXDs\BrokenLinks.mxd'
mxdObject = arcpy.mapping.MapDocument(mxdPath)
brokenLinks = arcpy.mapping.ListBrokenDataSources(mxdObject)
for layer in brokenLinks:
    oldPath, newPath = layerDic[layer.name]
    layer.findAndReplaceWorkspacePath(oldPath, newPath)
    mxdObject.save()
```

These solutions work well for individual map documents and layers. They can also be extended to folders full of MXDs by using the `glob.glob` method of the built-in `glob` module (which helps to generate a list of files that match a certain file extension) and the `os.path.join` method of the `os` module:

```
import arcpy, glob, os
oldPath = r'C:\Projects\OldData'
newPath = r'C:\Projects\Data'
```

```
folderPath = r'C:\Projects\MXDs'
mxdPathList = glob.glob(os.path.join(folderPath, '*.mxd'))
for path in mxdPathList:
    mxdObject = arcpy.mapping.MapDocument(mxdPath)
    mxdObject.findAndReplaceWorkspacePaths(oldPath,newPath)
    mxdObject.save()
```

Exporting to PDF from an MXD

Another important use of `arcpy.mapping` is to automatically export maps from map documents. The following code will export PDFs, but the module also supports the export of JPEGs and other formats. Using `arcpy.mapping` for this process avoids opening and exporting the MXDs:

> The output folder `C:\Projects\PDFs` must exist for this code to run correctly. While there are `os` module methods to check if a path exists (`os.path.exists`) and to create a folder (`os.mkdir`), in this code snippet the `arcpy.mapping.ExportToPDF` method will throw an exception if the input or output paths do not exist.

```
import arcpy, glob, os
mxdFolder = r'C:\Projects\MXDs'
pdfFolder = r'C:\Projects\PDFs'
mxdPathList = glob.glob(os.path.join(mxdFolder, '*.mxd'))
for mxdPath in mxdPathList:
    mxdObject = arcpy.mapping.MapDocument(mxdPath)
    arcpy.mapping.ExportToPDF(mxdObject, os.path.join(pdfFolder,
    os.path.basename(mxdPath.replace('mxd','pdf'))))
```

> Try turning this example code into a function that would accept the folder path as a parameter, or turning it into a script tool to create a custom map exporting tool. Use the `os` module to test if the file path exists with `os.path.exists`.

Automated map document production

The `arcpy.mapping` module includes important methods that will facilitate the automation of map document manipulation. These include the ability to add new layers or turn layers on and off within MXDs, the ability to adjust the scale of the data frame or move a data frame to focus on a specific region, and the ability to adjust text components of the map (such as titles or sub titles). These methods will be used as we continue our bus stop analysis.

Open up the MXD called `MapDocument1.mxd`. This represents our base map document, with layers and elements that we will adjust to our needs. It contains layers that we have used for our analysis, and other layers that fill out the map. There are also a number of text elements that will be automatically replaced by a script to fit the specific needs of each map. However, it does not do a good job of representing the results of the analysis as the census blocks that intersect the bus stop buffers overlap, making it hard to interpret the cartography.

The script will make it possible to produce dynamic map pages for each bus stop and the census blocks that are intersected with a buffer polygon geometry generated in memory around the stops:

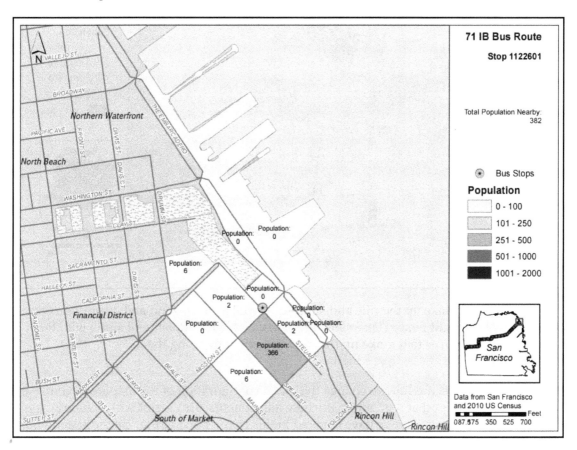

Here is an exported screenshot of the initial state of the map document:

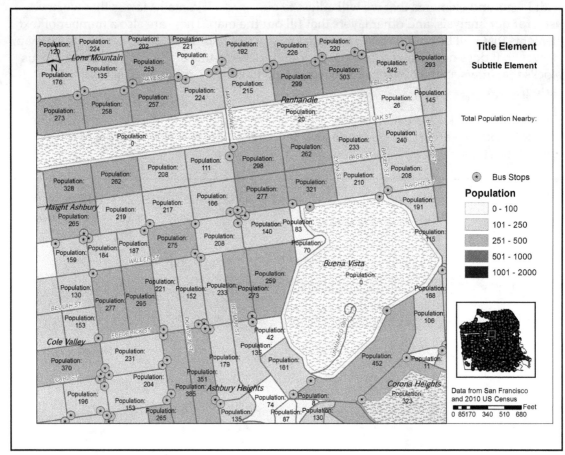

The text elements make up the title and subtitle, as well as the legend and attribution text at the bottom of the right pane. These elements are available for adjustment along with the layers and data sources that make up the map document by using the `arcpy.mapping.ListElements` method.

Now that we understand how to change the initial configuration of the map document, we will introduce a script that will automate these adjustments. This script will include a number of the concepts that we have covered in this chapter and earlier chapters, and will also introduce some new methods for map document adjustments that we will detail in the next section.

Check to make sure that the file paths for each layer are working, then close the MXD and open up the script `Chapter6_ExportMap.py`. Let's review it section by section to address what each part of the script is doing.

The variables

Within the script, a number of variables are first created to hold the string templates, the integer buffer distance, and the sql statement used to identify the bus line of interest:

```
#Import modules
import arcpy, os

#Assign local variables
mxdPath = r'C:\Projects\MXDs\MapDocument1.mxd'
outpathTemplate = r'C:\Projects\Map_{0}.pdf'
bufferDist = 400
whereCondition = "NAME = '71 IB' AND BUS_SIGNAG = 'Ferry Plaza'"
queryTemplate = "OBJECTID IN ({0})"
```

Connection to the map document

Use the `arcpy.mapping.MapDocument` method to connect to the MXD by passing the file path as a string. The object returned by this method (here assigned to the `mxdObject` variable) is passed to other functions as a parameter:

```
#Create the connection to the MXD
mxdObject = arcpy.mapping.MapDocument(mxdPath)
```

If this code is executed in the Python window of an open map document, don't pass an MXD file path to the `arcpy.mapping.MapDocument` method. Instead, use the keyword `"CURRENT"` to refer to the current open map document:

```
mxdObject = arcpy.mapping.MapDocument("CURRENT")
```

Data frames

Once the map document object has been created, the `Layers` data frame is selected from a list of data frames using the `ListDataFrames` method and passed to the variable called `dataFrame`. The `Inset` map (a small overview data frame under `Legend`) is also assigned to a variable:

```
dataFrame = arcpy.mapping.ListDataFrames(mxdObject, "Layers")[0]
insetFrame = arcpy.mapping.ListDataFrames(mxdObject, "Inset")[0]
```

Access the layers

Let's take a look at the following steps to access the layers:

1. Use the `arcpy.mapping.ListLayers` method to obtain a list of all layers within a specified `dataFrame` variable. Here, we are accessing the bus stop and the census block layers:

```
layersList = arcpy.mapping.ListLayers(mxdObject,"",dataFrame)
layerStops = layersList[0]
layerBlocks = layersList[3]
```

2. Get MXD layers from the `Inset` data frame and add a definition query to the inset bus stop layer to only show the stops of interest:

```
layersList = arcpy.mapping.ListLayers
(mxdObject,"",insetFrame)
layerStopsInset = layersList[0]
layerStopsInset.definitionQuery = whereCondition
```

The layout elements

The layout elements are returned as a list to the elements variable using the `ListLayoutElements` method. The layout elements include the various elements of the map document layout view such as the legend, the neat lines, the north arrow, the scale bar, and the text elements used as titles and descriptions. Unfortunately, there is no nice order to the list returned, as their position throughout the layout is undetermined. To access the text, elements that we would like to assign to a variable for later use must be identified using two properties of the element objects such as first `type` and then `text` (as we are adjusting objects with a `text` property). We want to discover and adjust the title, subtitle, and the total population elements, so a `for` loop is used to search through the list of elements:

```
elements = arcpy.mapping.ListLayoutElements(mxdObject)
for el in elements:
```

```
if el.type =="TEXT_ELEMENT":
   if el.text == 'Title Element':
       titleElement = el
   elif el.text == 'Subtitle Element':
       subTitleElement = el
   elif el.text == 'Total Population Nearby:':
       populationElement = el
       populationtext = populationElement.text
```

Generating a buffer from the bus stops feature class

All of the variables have been generated or assigned, so the next thing is to use
`SearchCursor` to search through the selected bus stops. For each bus stop, buffer objects
will be generated to find census blocks that intersect with these individual bus stops:

```
with arcpy.da.SearchCursor(layerStops,'SHAPE@','STOPID',
   'NAME','BUS_SIGNAG' ],whereCondition) as cursor:
   for row in cursor:
        stopPointGeometry = row[0]
        stopBuffer = stopPointGeometry.buffer(bufferDist)
```

For each row of data retrieved from the bus stops feature class, a number of attributes are
returned, contained in a tuple. The first of these, `row[0]`, is a `PointGeometry` object. This
object has a buffer method that is used to generate a buffer Polygon object in memory,
which is then assigned to the `stopBuffer` variable.

Intersecting the bus stop buffer and census blocks

To identify the census blocks intersecting with the buffer around each bus stop, the
ArcToolbox tool `SelectLayerByLocation` is used with the ArcPy method
`SelectLayerByLocation_management`. This tool will select the census block geometries
that intersect with the bus stop buffer.

A list called collector is created to collect block Object IDs, and then the buffer geometry
around the bus stop is assigned to a variable called `unionGeometry`. This geometry will be
used in union operations with all surrounding block geometries to create a representation of
the total "service area" of the bus stop. The population integer variable is used to find the
total population in the "service area" blocks. With each iteration the population variable is
adding each block's `POP10` value to itself to create a running total for the "service area":

```
arcpy.SelectLayerByLocation_management(layerBlocks, 'intersect',
   stopBuffer, "",  "NEW_SELECTION")
     collector = []
     unionGeometry = stopBuffer
```

```
population = 0
with arcpy.da.SearchCursor(layerBlocks,["OID@","SHAPE@","POP10"])
   as cursor:
   for brow in bcursor:
        collector.append(brow[0])
        unionGeometry = unionGeometry.union(brow[1])
        population = population + brow[2]
arcpy.SelectLayerByAttribute_management(layerBlocks,"CLEAR_SELECTION")
```

Finally, the layer selection is cleared to ensure that the next bus stop buffer will be intersected against the all census block geometries.

In the preceding code, the `for` loop is used to add the variable population to itself, plus a new value, to create a running tally of the total population around each stop. Adding a variable to itself, plus another variable, is so common that there is a shorthand for it in Python: += (there is also shorthand for subtracting: -=). This also works for strings. Here are examples using this shorthand, from an IDLE session:

```
# Plus Equals
>>> x = 1
>>> for num in range(1,5):
        x += num
>>> print x
11
# Minus Equals
>>> for num in range(1,5):
        x -= num
>>> print x
1
#String Addition
>>> a = ""
>>> for val in ['this', 'is','a','list']:
        a += val + " "
>>> print a
this is a list
```

Format a dynamic definition query

Use string addition to create a dynamic query by iterating through the list of object IDs (OIDs) returned by the second `SearchCursor`. This code checks to see if the list `collector` has collected any object IDs, using Python's built-in **len** function which accepts any iterable data type.

An integer variable, `counter`, and a string variable, `oidstring`, are assigned values; each will be used to tally data. Using a for loop, each object ID is converted into a string data type using the str function, and is added to the oidstring variable. The oidstring variable is added to itself, with a comma separating the object IDs. An `if` conditional checks the counter variable against the length of the list of object IDs; once it gets to the last object ID, no comma is added.

The string oidstring is then passed to the string queryTemplate's format method; the result is assigned to the variable `defQuery`. The newly created SQL, where conditional string `defQuery` is assigned to the census block layer's definition query method:

```
if len(collector) > 0:
    counter = 0
    oidstring = ""
    for oid in collector:
        if counter < len(collector)-1:
            oidstring = oidstring + str(oid) + ","
        else:
            oidstring = oidstring + str(oid)
        counter = counter + 1
    defQuery = queryTemplate.format(oidstring)
    layerBlocks.definitionQuery = defQuery
```

A Python built-in string method called **join** is useful for creating data strings from lists and tuples. It represents a simplified method of creating a new string from a list of strings, to avoid the need for iterating through the list and adding it to another string:

```
>>> alist = ['this', 'is','a','list']
>>> " ".join(alist)
'this is a list'

>>> ",".join(alist)
'this,is,a,list'
```

If every member of a list (or tuple) is a string, the list can be passed to join as a parameter. The data items in the list will be added together in list order, separated by whatever string character is used to call `join`. In the first example, the space string character is used to call join; the result is separated by spaces. In the second, a comma is used, and the result is a string with the list values separated by commas.

Updating the layout elements

Now that the data has been manipulated, the next thing we can do is to update the layout elements. This includes layer properties that will affect the legend, the extent of the data frame, and the text elements:

```
dataFrame.extent = unionGeometry.extent
dataFrame.scale = dataFrame.scale * 1.1

populationElement.text = populationtext + "\n" + str(population)
titleElement.text = row[2] + " Bus Route"
subTitleElement.text = "Stop " + str(row[1])

arcpy.RefreshActiveView()
```

The data frame extent is adjusted by assigning the variable to the extent of the unionGeometry, an `arcpy.Extent` object with four parameters, Xmin, Ymin, Xmax, and Ymax. Once the data frame's extent is adjusted, the scale of the data frame is then multiplied to include more of the surrounding area (backing the data frame out).

The population text for the bus stops layer is adjusted using the `populationElement.text` property, and assigned to the variable `population` to reflect the total number of people within the census blocks served by the current bus stop. The text elements are updated using their `text` property to show the current route and bus stop. Finally, the `RefreshActiveView` method is used to ensure that the map document window is correctly updated to the new extent.

Exporting the adjusted map to PDF

The final step is to pass the newly adjusted map document object to the `ExportToPDF` method of the ArcPy. This method requires two parameters, such as the map document object and a string that represents the file path of the PDF:

```
outpath = outpathTemplate.format(str(row[1]))
print outpath

arcpy.mapping.ExportToPDF(mxdObject,outpath)

#Clean up for the next iteration
layerStops.definitionQuery = ""
layerBlocks.definitionQuery = ""
```

The PDF file path string is generated from the `pdfFolder` string template and the id of the bus stop, along with the object `id` and the file extension `.pdf`. Once the PDF file path string and the map document object represented by the variable `mxdObject` are passed to the `ExportToPDF` method, the PDF will be generated. The text elements are then reset and the view is refreshed to ensure that the map document will be ready for the next iteration of the script.

The Results - Dynamic maps

Here is an example of the output:

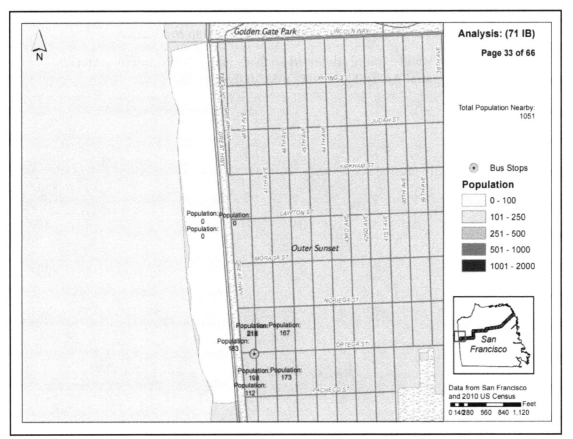

The maps generated by the script show each bus stop at the center, surrounded by the symbolized census blocks with which the buffer intersects. The title, subtitle, and the legend have been adjusted to indicate the bus stop depicted in the map. With ArcPy, we are now in control of both the parts of geospatial analysis, such as the analysis itself and the cartographic production depicting the result of the output.

Summary

In this chapter, `arcpy.mapping` was introduced and used to control the elements of map documents that need to be adjusted to create custom maps. File path adjustments were covered using a few methods. Layer objects and some of their properties and methods were explored. The Bus Stop analysis was expanded to include map production.

In the next chapter, we will explore advanced analysis using ArcPy and two ArcGIS extensions such as **Spatial Analyst** and **Network Analyst**.

7
Advanced Analysis Topics

ArcGIS for Desktop extensions benefit from the power of Python and ArcPy. In this chapter we'll explore two important examples of extensions, the Network Analyst and Spatial Analyst modules. Each of these modules have ArcPy wrapper tools to access ArcToolbox. They also have ArcPy access modules for improved control of available tools, methods, and properties. Using the extensions and ArcPy scripting, advanced analysis workflows such as modeling traffic or planning bus routes using a streets dataset can be automated.

This chapter will cover the following topics:

- Creating a simple network dataset
- Checking out the extensions
- The ArcPy Network Analyst module
- The ArcPy Spatial Analyst module

Using Network Analyst

ESRI's Network Analyst extension is a powerful tool for enabling routing and network connectivity functionality within ArcGIS. The extension is commonly used for street routing and finding the best path between two points along a road network. The route can be constrained by a number of impedence factors, such as traffic or left turns, to best model the reality of road travel. Similar analyses can be run using other types of networks, such as water pipe networks or electrical networks.

To use the Network Analyst extension, the ArcGIS for Desktop advanced license is required. In ArcCatalog or ArcMap, click on the **Customize** menu, and select **Extensions**. Once the **Extensions** menu is open, click on the checkbox next to **Turn on the Network Analyst extension**.

Creating a network dataset

The first step in using a network dataset is to create one within a feature dataset. To do so, we will generate a feature dataset to hold the data of interest. Right-click on the **File Geodatabase**, which houses the bus stop data, and select **New.** Then select the **Feature Dataset** option from the **New** menu. Name it Chapter7Results, and click on the **Next** button.

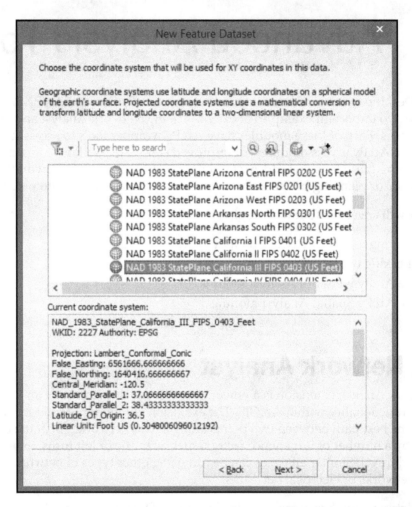

Next, select **Spatial Reference System (spatial reference system)**. In this case, we will be using the spatial reference system of the local State Plane zone for San Francisco. It is a Projected coordinate system, so select that folder, then click on the State Plane folder. Once it is opened, select the folder called `NAD 1983(US Feet)`. From the available reference systems, select `NAD 1983 StatePlane California III FIPS 0403 (US Feet)`. Push **Next** to go to the next menu.

Click on the `Vertical Coordinate Systems` folder, and select the `North America` folder. Select the **NAVD 1988 US survey feet** (North American Vertical Datum of 1988 in feet) option. This will make it possible to have the vertical and horizontal linear units in the same measurement system. Click **Next** to go to the next menu.

The tolerances on the next page are also very important, but we will not cover them in detail here. Accept the defaults, and then click on the **Finish** button to finalize the feature dataset.

 This coordinate system is known as **2227** in the **Well Known ID (WKID)** or **European Petroleum Survey Group** (**EPSG**) systems. More information about these codes is available at `http://spatialreference.org`, a website used to find the thousands of spatial reference systems used throughout the world. They are useful for querying published services.

Importing the datasets

Import the bus stops, streets, and bus routes feature classes into the `Chapter7Results` Feature Dataset. Right-click on the dataset, and select **Import | Feature Class (Single)**. Add the feature classes one by one to give them a new name, which will keep them separated from the versions contained within the **SanFrancisco Feature Dataset**. Importing the feature classes will ensure they are in the correct spatial reference system so a network dataset can be created.

Creating the network dataset

Now that we have a data container, we can create a network dataset from the streets feature class. Right-click on the **Chapter7Results** feature dataset, and select from the list **New** | **Network Dataset**, as shown in the following screenshot:

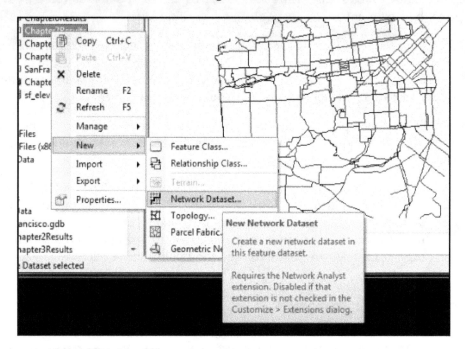

Call the network dataset Street_Network, and click **Next**. Select the Streets feature class as the class that will participate in the network dataset, and push **Next** to move to the next menu. Select **Global Turns** to model turns within the network. In the next menu, use the default connectivity settings. Then accept the **Using Z Coordinate Values from Geometry** setting. Accept the default cost restriction and driving directions settings, and finally, push finish to generate the network dataset. Then, build the network dataset using the **Final** menu. The network dataset is ready to be used.

Accessing the network dataset using ArcPy

Now that the necessary setup is complete, the Street_Network network dataset can be added to a script for use in generating routes. Because this is a simple analysis, the only impedance value to be used will be the length of the street segments. Through the use of the SearchCursor method, the PointGeometry objects from the bus stops can be accessed and added as locations to be searched as follows:

```
import arcpy

arcpy.CheckOutExtension("Network")
busStops = r'C:\Projects\SanFrancisco.gdb\Chapter7Results\BusStops'
networkDataset =
r'C:\Projects\SanFrancisco.gdb\Chapter7Results\street_network'
networkLayer = "streetRoute"
impedance = "Length"
routeLayerFile = "C:\Projects\Layer\{0}.lyr".format(networkLayer)

arcpy.MakeRouteLayer_na(networkDataset, networkLayer, impedance)
print 'layer created'
sql = "NAME = '71 IB' AND BUS_SIGNAG = 'Ferry Plaza'"
with arcpy.da.SearchCursor(busStops,['SHAPE@', 'STOPID'],sql) as cursor:
    for row in cursor:
        stopShape = row[0]
        print row[1]
        arcpy.AddLocations_na(networkLayer,'Stops',stopShape, "", "")
arcpy.Solve_na(networkLayer,"SKIP")
arcpy.SaveToLayerFile_management(networkLayer, routeLayerFile,"RELATIVE")
print 'finished'
```

Breaking down the script

Let's dissect the preceding script, which, once finished, will generate a layer file containing the added *stops*, and the *routes* along streets to best get from the origin stop to the destination stop.

The script begins by importing the arcpy module, and the next line allows us to use the Network Analyst extension:

```
arcpy.CheckOutExtension("Network")
```

ArcPy's `CheckOutExtension` method checks out the Network Analyst extension when the correct keyword is passed to the method as a parameter. Once it has been invoked, the tools of the extension can be called and executed in the script.

Once the bus stops feature class and the `street_network` network dataset have been assigned to variables, they can then be passed to `ArcPy's MakeRouteLayer_na` method along with a variable representing the impedance value as follows:

```
arcpy.MakeRouteLayer_na(networkDataset, networkLayer, impedance)
```

The `MakeRouteLayer_na` tool produces a *Route Layer* in memory. This blank layer needs to be populated with stops to produce the route(s) between them. For this purpose, we need a `SearchCursor` to access the `PointGeometry` objects, and an SQL statement that will limit the returned results to the line of interest.

```
sql = "NAME = '71 IB' AND BUS_SIGNAG = 'Ferry Plaza'"
with arcpy.da.SearchCursor(busStops,['SHAPE@', 'STOPID'],sql) as cursor:
for row in cursor:
    stopShape = row[0]
    print row[1]
    arcpy.AddLocations_na(networkLayer,'Stops', stopShape,"", "")
```

The `SearchCursor` will allow the `Stops` sublayer of the layer produced by the `MakeRouteLayer` tool to be populated when used in conjunction with the `AddLocations` tool. Once populated, the Route Layer can be passed to the `Solve` tool to find the routes between the points of interest. Again, the routes are solved based on finding the lowest *impedance* between the two points. In this example, the only impedance is segment length, but it could be traffic, elevation, or other restriction types, if that data is available.

```
arcpy.Solve_na(networkLayer, "SKIP")
arcpy.SaveToLayerFile_management(networkLayer, routeLayerFile, "RELATIVE")
```

The final result is a layer file, which is written to disk using the `SaveToLayerFile` tool, is shown in the following screenshot:

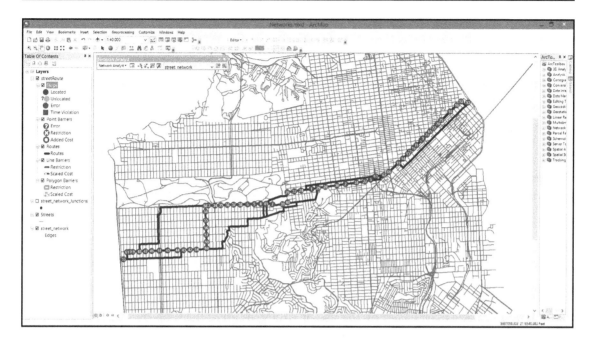

The Network Analyst module

In an effort to make the use of the Network Analyst extension more *Pythonic*, the newer **Network Analyst (NA)** module adjusts how the methods that correspond to the **ArcToolbox** Network Analyst tools are accessed. Instead of calling the tools directly from ArcPy, the tools are now methods of the NA module. Renaming Network Analyst toolset reduces confusion and makes it easier to remember the name of the method. See the differences as follows:

```
import arcpy
arcpy.CheckOutExtension("Network")
busStops = r'C:\Projects\SanFrancisco.gdb\SanFrancisco\Bus_Stops'
networkDataset =
  r'C:\Projects\SanFrancisco.gdb\Chapter7Results\street_network'
networkLayer = "streetRoute"
impedance = "Length"
routeLayerFile = "C:\Projects\Layer{0}_2.lyr".format(networkLayer)
arcpy.na.MakeRouteLayer(networkDataset, networkLayer, impedance)
print 'layer created'
sql = "NAME = '71 IB' AND BUS_SIGNAG = 'Ferry Plaza'"
with arcpy.da.SearchCursor(busStops,['SHAPE@','STOPID'],sql) as cursor:
    for row in cursor:
```

```
        stopShape = row[0]
        print row[1]
        arcpy.na.AddLocations(networkLayer,'Stops',stopShape, "", "")
arcpy.na.Solve(networkLayer,"SKIP")
arcpy.management.SaveToLayerFile(networkLayer, routeLayerFile,"RELATIVE")
print 'finished'
```

The tool will produce the same layer output as the original script, but the reorganization of the Network Analyst tools in the NA module has made the code more logical. For instance, it makes more sense to call the `Solve` tool using `arcpy.na.Solve()` instead of `arcpy.Solve_na()`, as it reinforces that `Solve` is a method of the NA module. As ArcPy continues to be developed, I expect more *Pythonic* code reorganization to occur.

Accessing the Spatial Analyst extension

The Spatial Analyst extension is very important for performing analysis on both raster and vector datasets, but it is generally used to perform surface analyses and raster math. These operations are made even easier by the use of ArcPy, as all of the tools available in the Spatial Analyst toolbox are exposed with the Spatial Analyst access module. This includes the raster calculator tools, making map algebra easy by using the tools and operators in simple expressions.

Adding elevation to the bus stops

The elevation raster, `sf_elevation`, has been downloaded from **National Oceanic and Atmospheric Administration** (**NOAA**), and added to the `Geodatabase` file. However, it covers the city of San Francisco, and we should write a script to only extract an area of the neighborhood of San Francisco, as it will reduce the time needed to run our scripts. We'll use an SQL statement as the where clause to limit the results to the **South of Market** (**SoMa**) neighborhood. To do so, let's take advantage of a Search Cursor and the Spatial Analyst access module's `ExtractByPolygon` property:

```
import arcpy
arcpy.CheckOutExtension("Spatial")
busStops = r'C:\Projects\SanFrancisco.gdb\SanFrancisco\Bus_Stops'
sanFranciscoHoods =
  r'C:\Projects\SanFrancisco.gdb\SanFrancisco\SFFind_Neighborhoods'
sfElevation = r'C:\Projects\SanFrancisco.gdb\sf_elevation'
somaGeometry = []
sql = "name = 'South of Market'"
with arcpy.da.SearchCursor(sanFranciscoHoods,['SHAPE@XY'],sql,None,
  True) as cursor:
```

```
    for row in cursor:
        X = row[0][0]
        Y = row[0][1]
        somaGeometry.append(arcpy.Point(X,Y))
    somaElev = arcpy.sa.ExtractByPolygon(sfElevation,somaGeometry,
                                "INSIDE")
    somaOutPath = sfElevation.replace('sf_elevation','SOMA_elev')
    somaElev.save(somaOutPath)
    print 'extraction finished'
```

The `ExtractByPolygon` method is a bit misleading, as it does not accept a Polygon object as a parameter. Instead, it requires a list of point objects, which represent the vertices of the area that we want to extract. As the Search Cursor iterates through the neighborhoods dataset, a `Polygon` object is returned by the cursor. Fortunately, the Search Cursor has a final parameter, which we have not yet explored, which allows us to extract the individual points or vertices that make up the SoMa neighborhood polygon. By setting the Search Cursor's optional `Explode to Points` parameter (which converts Polygon objects into coordinate pairs for each vertex) to true, Point objects can be generated by passing the `X` and `Y` values of each returned vertex to the `arcpy.Point` method. These Point objects are appended to the list `somaGeometry`, and then passed to the Spatial Analyst access module's `ExtractByPolygon` method.

Note that passing a Polygon Object instead of Point Objects will return an error.

Using Map algebra to generate elevation in feet

We now have a raster to use for extracting elevation values. However, both the original raster and the generated SoMa neighborhood raster contain elevation values in meters, and it would be better to convert them to feet to keep them consistent with the projection of the bus stops. Let's use raster math and the `Times` method to convert the values from meters to feet:

```
    somaOutPath = sfElevation.replace('sf_elevation','SOMA_elev')
    outTimes = arcpy.sa.Times(somaOutPath, 3.28084)
    somaFeetOutPath = sfElevation.replace('sf_elevation','SOMA_feet')
    outTimes.save(somaFeetOutPath)
```

The `Times` method generates a new raster to glean the elevations values we need for the bus stops of interest.

Adding in the bus stops and getting elevation values

Now that we have generated a raster that we can use to find elevation values in feet, we need to add in a new `arcpy.sa` method to generate the points. The `ExtractValuesToPoints` method will generate a new bus stops feature class with a new field that holds the elevation values.

```
arcpy.MakeFeatureLayer_management(busStops, 'soma_stops')
with arcpy.da.SearchCursor(sanFranciscoHoods,['SHAPE@'],sql) as cursor:
    for row in cursor:
        somaPoly = row[0]
        arcpy.SelectLayerByLocation_management("soma_stops",
        "INTERSECT", somaPoly)
outStops = r'C:\Projects\SanFrancisco.gdb\Chapter7Results\SoMaStops'
  arcpy.sa.ExtractValuesToPoints("soma_stops", somaOutFeet,
  outStops,"INTERPOLATE", "VALUE_ONLY")
print 'points generated'
```

The final result

We have now produced a subset feature class of the bus stops that have the elevation values added as a field. This process could be repeated for the entire city, one hood at a time, or it could be performed with the original elevation raster on the entire bus stops feature class to generate a value for each stop.

```
import arcpy
arcpy.CheckOutExtension("Spatial")
arcpy.env.overwriteOutput = True
busStops = r'C:\Projects\SanFrancisco.gdb\SanFrancisco\Bus_Stops'
sanFranciscoHoods =
  r'C:\Projects\SanFrancisco.gdb\SanFrancisco\SFFind_Neighborhoods'
sfElevation = r'C:\Projects\SanFrancisco.gdb\sf_elevation'
somaGeometry = []
sql = "name = 'South of Market'"
with arcpy.da.SearchCursor(sanFranciscoHoods,['SHAPE@XY'], sql,None,
  True) as cursor:
    for row in cursor:
        X = row[0][0]
        Y = row[0][1]
        somaGeometry.append(arcpy.Point(X,Y))
        somaElev = arcpy.sa.ExtractByPolygon(sfElevation, somaGeometry,
                                             "INSIDE")
        somaOutput = sfElevation.replace('sf_elevation','SOMA_elev')
        somaElev.save(somaOutput)
print 'extraction finished'
```

```
somaOutput = sfElevation.replace('sf_elevation','SOMA_elev')
outTimes = arcpy.sa.Times(somaOutput, 3.28084)
somaOutFeet = sfElevation.replace('sf_elevation','SOMA_feet')
outTimes = arcpy.sa.Times(somaOutput, 3.28084)
outTimes.save(somaOutFeet)
print 'conversion complete'
arcpy.MakeFeatureLayer_management(busStops, 'soma_stops')
with arcpy.da.SearchCursor(sanFranciscoHoods, ['SHAPE@'],sql) as cursor:
    for row in cursor:
        somaPoly = row[0]
        arcpy.MakeFeatureLayer_management(busStops, 'soma_stops')
        arcpy.SelectLayerByLocation_management("soma_stops",
                                    "INTERSECT", somaPoly)
outStops = r'C:\Projects\SanFrancisco.gdb\Chapter7Results\SoMaStops'
arcpy.sa.ExtractValuesToPoints("soma_stops",
somaOutFeet,outStops,"INTERPOLATE","VALUE_ONLY")
print 'points generated'
```

This preceding script demonstrates the value of accessing the advanced extensions in ArcPy, and combining them with Search Cursors and Geometry objects. The script could be taken even further by adding in a Search Cursor to look through the outStops dataset and exporting the results to a spreadsheet, or even adding a new field to the original bus stops dataset to populate with the elevation values. It could even be used as impedance values to be entered into a Network Analyst extension analysis, a fun coding task that I hope you will attempt.

Summary

In this chapter, we covered the basics of using common ArcGIS for Desktop Advanced extensions within ArcPy, with a focus on the Network Analyst access and Spatial Analyst access modules. We explored how to generate a network, and how to create network paths using ArcPy. We also explored how to access Spatial Analyst tools, and to use them in conjunction with Search Cursors to work with rasters and vectors for spatial analysis.

In the next chapter, we will transition to web mapping using ArcGIS Online, and explore how to use Python to automate the process.

8
Introduction to ArcGIS Online

In this chapter, we will discuss the use of ArcGIS Online to create web maps and share geospatial data. ArcGIS Online, created by Esri, is built to easily integrate web mapping into any production ArcGIS environment. By adding the ability to publish data to the cloud directly from an MXD, Esri has made it easy to share analysis results with the public or with a specific group of users.

This chapter will cover the following topics:

- The basics of ArcGIS Online
- Signing up for an ArcGIS online account
- Feature services and layers
- Publishing data from an MXD

ArcGIS Online

ArcGIS Online (**AGOL**) is one of Esri's web-mapping platforms along with the ArcGIS API for JavaScript, and Portal for ArcGIS. It is a complete platform, which includes cloud-based data storage, management, analysis, and sharing capabilities. AGOL has a free and a paid component, and storage and analysis beyond the free tier are managed using a credit system of Pay-As-You-Go tokens. Some contracts with Esri include access to ArcGIS Online and a set number of tokens to use, so check with the ArcGIS administrator within your company or organization to find out what you can do with an organization's AGOL site.

Within AGOL, there are a number of existing data resources published by other users, which can be incorporated into new web maps. With the free tier and subscription accounts, datasets and maps can be made available to the public so that other users can incorporate any data you publish, or they can be made private using sharing permissions, which will be discussed in this chapter.

Signing up for an account

To sign up for a free ArcGIS Online, you must create a public Esri account. With this account, you will be able to load data into the Cloud, and share it with the public in a custom web map. If you don't have an account yet, follow the following steps to create one:

1. Go to `https://arcgis.com/home/signin.html`:

2. Click on the **CREATE A PUBLIC ACCOUNT** option as shown in the following screenshot:

3. Use either a Facebook or Google account to log in, or to create a new account:

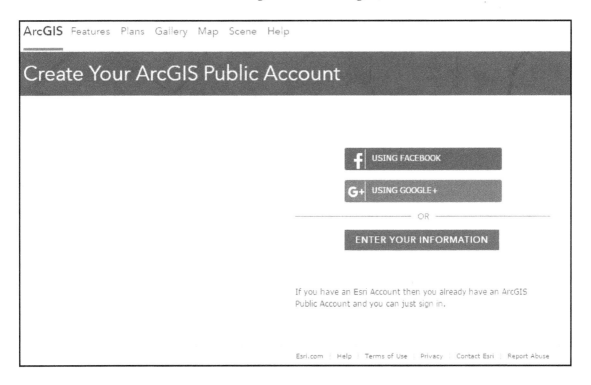

4. Fill in the usual account details including a username, password, e-mail, and a security question, as seen in this screenshot:

5. Push the **CREATE MY ACCOUNT** button after accepting the terms of use.

 If your company or employer has a subscription to ArcGIS Online included in the contract with Esri, then an employer-specific login must be created and approved by your ArcGIS Online administrator.

Exploring the interface

The AGOL interface has a number of tabs with functionality related to the different components of managing the online storage, and sharing of data and maps. Each of these should be explored to understand their use:

The My Organization tab

The **My Organization** tab organizes and controls the approved users of your AGOL site. With a free tier account, there can only be one user. Within organizational sites, this tab is where the approved ArcGIS Online administrator will assign permissions and add users:

The My Content tab

The **My Content** tab is the main tab for organizing layers, services, maps, and applications built in AGOL. All of the hosted data and services registered with the site will be listed and organized by title, type, last modified date, and sharing permissions, as shown here:

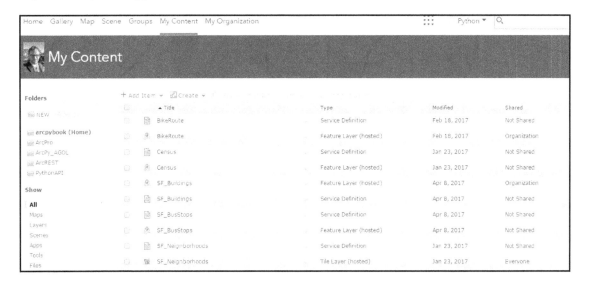

Within the **My Content** tab, the **Add Item** and **Create** drop-down options are used when new data is uploaded, or when new maps and mapping applications can be created. Folders (on the left) can be created to organize data and maps into logical groups. By default, all layers and maps are created in the Home folder.

The Add Item option

Click on the **Add Item** option to access a menu that allows for uploading data from the web or from a computer. Services published using ArcGIS for server or another web mapping server, hosted layers, and even shapefiles and CSVs can be added to AGOL using this button. We will discuss the most popular method, publishing data from an MXD, as shown in the following screenshot. However, there are other methods to upload and visualize data on an AGOL map:

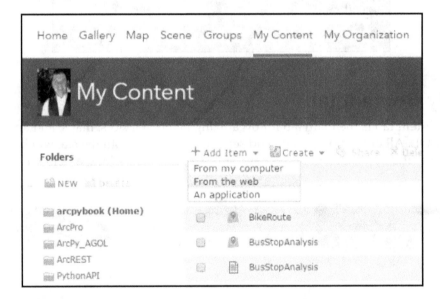

 Hosted feature layers can consume credits depending on the account type and size of the layer.

Features from services

Perform the following few steps to add an ArcGIS service:

1. To add an ArcGIS service or another web map service published using a web map server, click on **Add Item**, and then select **From the web**.
2. Specify the server used to publish the service from the list of radio buttons, and then supply the URL of the service, a title, and tags (which can be anything you chose) to make the data available.

> If the service is loaded from the ArcGIS server, there is no limit to the number of features loaded in an AGOL web map. Performance is only limited by the map server publishing the map service.

Features from files

There are a few limitations to add data files from a computer:

- Shapefiles, which are made up of at least three files, (`.shp`, `.dbf`, and `.shx` are required, and others such as `.prj` are optional, but encouraged) must be zipped
- Spreadsheets with point data must be in the comma-separated value (CSV) format, and must have the location data in different fields (x and y or latitude and longitude)

Taking into consideration the preceding mentioned limitations, perform the following steps to add data files from a computer:

1. Click on the **Add Item** button, and use the **Chose File** button to select the zipped folder or the CSV.
2. Supply a title and a tag for search discovery, and click **Add Item**.

> Feature layers are best limited to 1,000 rows per dataset. Any data larger than this will make the layer sluggish and slow within a web map.

The Create tab

The **Create** drop-down menu has a number of powerful tools. It links to the **Map** and **Scene** interfaces, which can also be accessed using the respective tabs at the top of the page. It also allows the user to create a feature layer or web map data layer from an existing template, or from data already hosted on the AGOL site:

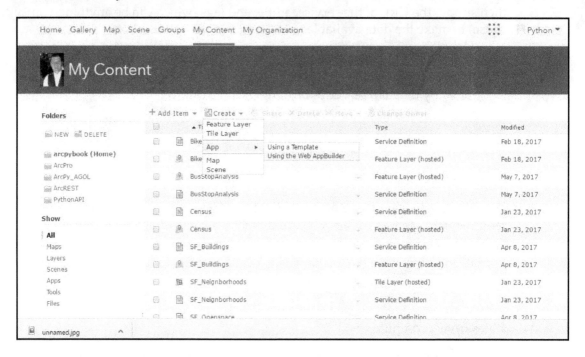

Importantly, web mapping applications can be created using the **Create** menu. While not covered here, there are a number of powerful and useful pre-built templates for creating web mapping applications within minutes. The Esri WebApp Builder tool is also very powerful, and allows for custom web mapping applications to be built with no coding required (not that we're afraid of getting our hands messy with code here!).

As a weekend challenge, try creating a story map using one of the pre-built templates and your own data. It is accessed by navigating to **Create** | **App** | **Using a Template**.

The Groups tab

The **Groups** tab allows AGOL administrators to group together specific layers and maps, and to control access to those resources. It makes it easy to assign permissions to resources for project members of each AGOL site hosted on the AGOL account.

To create a group, click on the **CREATE A GROUP** link below the **My Groups** banner. It links to a page where the user will supply the details for their group, including whether the group is public or private. Fill out the page to create a group, and be sure to select the correct group permissions. They can be edited later, but it's important to protect the security of hosted data when inviting others to view it:

The Map and Scene tabs

These tabs take the user to an environment where a web map or 3D scene can be created. Explore the **Map** tab, using data published from an ArcGIS for Desktop MXD and hosted in AGOL, along with one of the free Esri basemaps. To create 3D scenes, ArcGIS Pro publishes scene layers, and then shares them with AGOL.

These maps and scenes can be created, saved, and shared on their own, or they can be used as the basis for web mapping applications built using WebApp Builder, pre-built templates, Operations Dashboard, or even the ArcGIS API for JavaScript.

Click on each tab, and explore the interfaces to get comfortable with them. Each interface has loads of data management and cartographic functionality, which influences the look and feel of the maps or scenes produced. Basemaps and hosted data layers can be added, popups can be configured, and data layers can be styled, all within these tabs. While exploring these completely is out of the scope of this book, be sure to spend time exploring them.

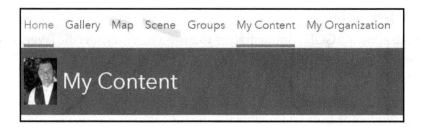

Publishing from an MXD

Publishing a layer or layers from an MXD is a great way to use the familiar cartographic options within ArcGIS for Desktop to style the layers. Open an MXD within ArcGIS Online, and load the `Bus Stops` and `San Francisco` feature classes from the SanFrancisco geodatabase to get started:

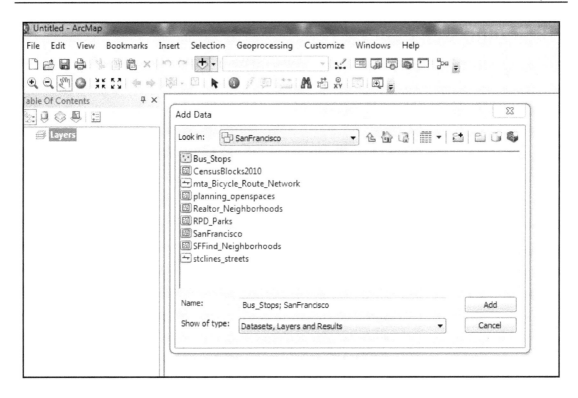

Styling the layers

Once the layers are loaded, double-click on the Bus Stop layer in the Table of Contents, and open the Layer Properties menu to style the layer. Then follow these steps:

1. Open the Symbology tab in the Layer Properties menu.
2. Click on the Symbol button to open the Symbol Selector.
3. Make the symbol size 6.
4. Select a color from the color drop-down interface by clicking on the color menu button.
5. Push the OK button at the bottom of the Symbol Selector.
6. Chose a color for the San Francisco layer using the same process.

These style choices will be carried with the layers once they are published to AGOL.

Publishing the layers

Now that the layers have been styled to meet our needs, we need to publish the layers and share them with AGOL. To do so, we'll use the **File** menu to sign into AGOL, and to share the layers:

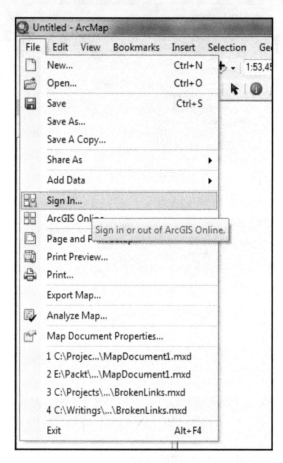

Open the **File** menu, and select **Sign In...**. A menu will appear with entries for the AGOL username and password created earlier in the chapter:

The Share As menu

Once signed in, the next step is to use the **Share As** menu. Click on the **File** menu, and locate the **Share As** menu item. Select **Service...** from the sub-menu as seen in the following screenshot:

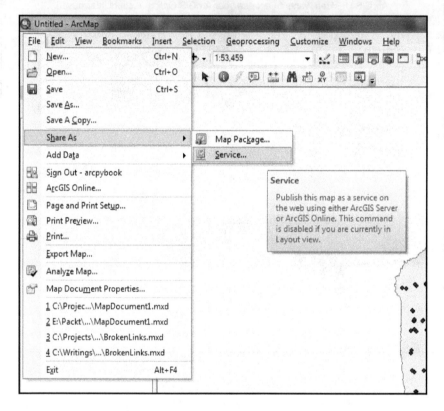

Once the Service menu appears, select **Publish a service**, and click **Next**:

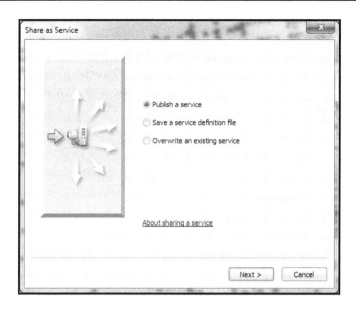

On the next menu page, select the connection from the list at the top. If you have registered an ArcGIS Server instance, it will appear here as well. In this case, select **My Hosted Services**, and give the service a name:

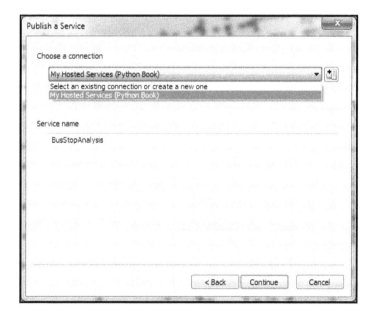

Click **Continue** to go to the Service Editor.

Service Editor

The Service Editor is used for a number of administrative tasks before publishing the service. Specifying the type of service to be published, the feature access capabilities, sharing permissions, and an item description are all done here. It also provides an analyzer to test for any issues that may prevent it from being shared:

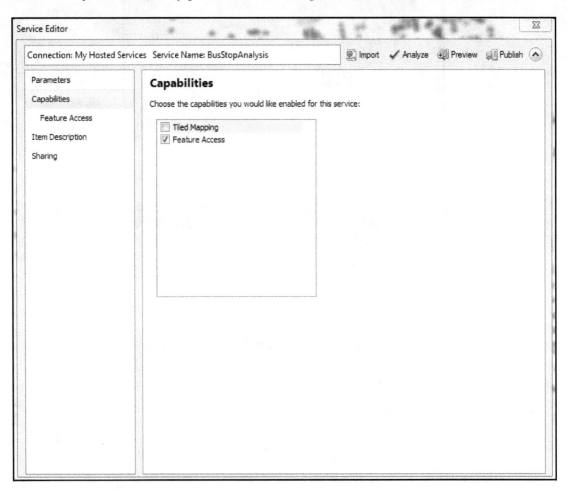

In the **Parameters** tab on the left, the maximum number of records returned (usually 1,000 per request) is adjustable. The type of service to be published is selected in the **Capabilities** tab. The description and tags and credits for the layer are established in the **Item Description** tab, and **Permissions** tab are controlled in the **Sharing** tab. Explore each of these to understand their capabilities and limitations better.

The Item Description option

Make sure that the **Capabilities** tab is set to a **Feature Access** service type, then open the **Item Description** tab, and add a short description of the service, along with a tag such as San Francisco or Bus Stop Analysis. Add credits for attribution of the data as seen in this screenshot:

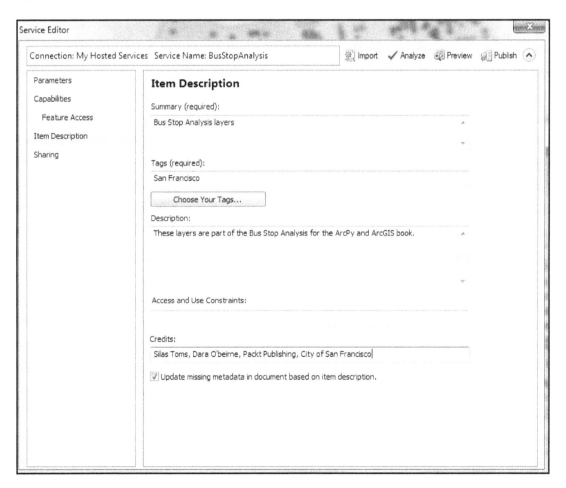

Analyze

Once these are filled out, and the **Sharing** tab permissions are set to the user's liking, it's time to analyze the service, and assess its readiness for publication. Let's take a look at the following steps to see how it is done:

1. Click on the **Analyze** button at the top. An interface appears within the MXD, which lists any errors or warnings about the layers as shown in the following screenshot:

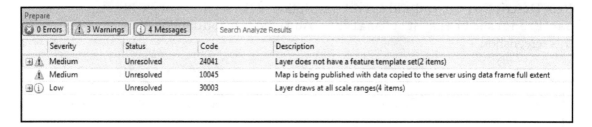

	Severity	Status	Code	Description
⊞ ⚠	Medium	Unresolved	24041	Layer does not have a feature template set(2 items)
⚠	Medium	Unresolved	10045	Map is being published with data copied to the server using data frame full extent
⊞ ⓘ	Low	Unresolved	30003	Layer draws at all scale ranges(4 items)

2. If there were errors, they must be resolved before the layer can be published. Click on each error to resolve it.

3. The analysis results seen in the preceding screenshot indicate that there are no errors, but there are three warnings and a few messages. These can be explored by clicking on each line in the table. Errors can similarly be resolved using this table interface.

4. Once the errors are resolved, and the warnings are resolved or ignored, it is safe to publish the layers as a service. Push **Publish** in the upper-right corner of the screen. A status bar window will appear to inform you of the progress, as shown in the following screenshot:

After creating the service definition and then loading the data into the cloud, the service will be published, and will appear in the **My Content** tab of ArcGIS Online:

Updates

Updating the service is as simple as updating the data within the MXD used to publish the service, and going through the process outlined previously. Plan for the storage of service MXDs, as it will make it easier to republish layers, and retain the cartographic representation.

Work to automate the data management for the MXDs using Python. Whether updating an SDE database or a File Geodatabase, ArcPy, and the ArcToolbox will make it easier to control the data source that is published to the cloud.

Developer account

Now that the process of signing up to ArcGIS Online is understood, let's explore the next level of ArcGIS Online web map development. The same account can be converted to a developer account by performing the following steps:

1. Navigate to `https://developers.arcgis.com/sign-in/`.
2. Supply your username and password.
3. Click the **SIGN IN** button:

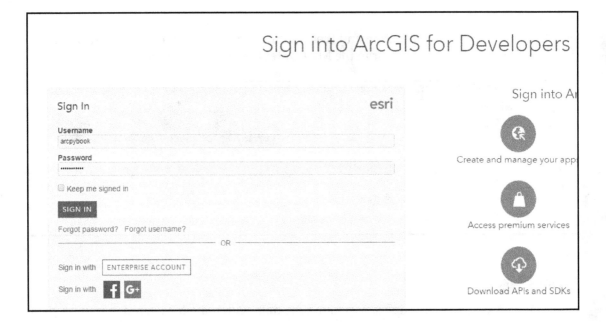

With this account, a developer dashboard layout is used. It offers easy download of the Esri APIs and SDKs to enable web map development and data management. It lists layers and applications with a guided application development section. The free developer account offers a number of development credits (50 at publishing time), which is trackable on the dynamic graph. This account will be used for the next few chapters:

Summary

In this chapter, we explored the basics of ArcGIS Online, and created a public account, which allows us to publish data from ArcGIS for Desktop, or to load data into the Cloud. With this account and the published services, we can share data, create web maps, and even create complex web mapping applications without coding.

In the next chapter, we will explore the use of Python for ArcGIS Online data management. We will use the ArcGIS API for Python, which allows us to automate the process outlined earlier, as well as many other common data management tasks.

9
ArcPy and ArcGIS Online

Now that we have reviewed the basics of ArcGIS Online, we will go through some more advanced functionalities and begin to build the foundation to interact with ArcGIS Online hosted feature services using Python. In order to program with Python and ArcGIS Online, we will need to first understand how the data is formatted and the methods used to access that data programmatically.

This chapter will cover the following topics:

- ArcGIS Online REST services
- URL parameters
- Using ArcGIS feature sets
- ArcGIS Online tokens
- Updating a feature class from an AGOL-hosted feature service

ArcGIS Online REST services

Layers and maps in ArcGIS Online are exposed through the Web using a communication standard known as **Representational State Transfer** (**REST**). These web services are commonly referred to as REST Services. REST is a lightweight communication format between the producer and the consumer, which makes it extremely popular among cloud-based Application Programming Interfaces, or APIs, such as the one produced by ESRI. When a producer exposes their web services using the REST architecture, they are called RESTful APIs or a REST API. In our case, we refer to the maps and layers exposed through ArcGIS Online using REST as the ArcGIS REST API.

With the ArcGIS REST API, you can perform multiple queries, share data within your organization, expose data from your organization to the public, and configure parts of your internal ArcGIS infrastructure, such as ArcGIS Server and Portal. When discussing the ArcGIS REST API, it is important to understand that the API applies to ArcGIS Server, ArcGIS Online, and Portal for ArcGIS. Through the API, a user can consume various types of services, including feature services, map services, geocoding services, image services, and geoprocessing services, among others. In order to consume these services, you will need to know the URL for the service. The URL for the service can either point directly to your ArcGIS Server or to an ArcGIS Online account.

Exploring ArcGIS REST services

Using the ArcGIS Online account you set up in Chapter 8, *Introduction to ArcGIS Online*, let's publish the San Francisco Bus Stops feature class to your account and call it **BusStops**. Once complete, you should see the hosted feature layer under **My Content**, as shown here:

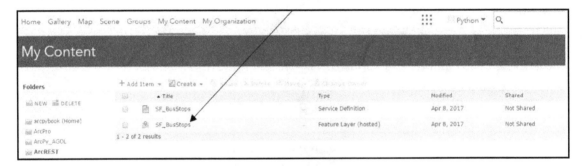

To view the feature layer and access the service, you need to click on the **SF_BusStops** link under the title. Next, you should see the service definition, as shown in the following screenshot:

Next, in order to access the rest service for this feature layer, scroll down and click on **Service URL**:

The rest endpoint for this hosted feature layer should look like this:

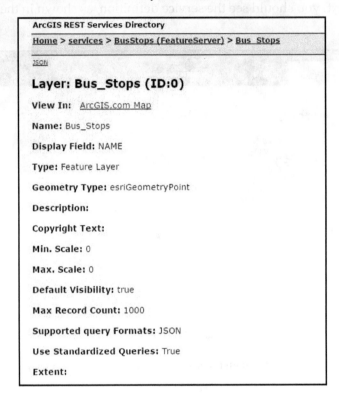

In this REST service definition, you should see information regarding the feature that was published. You'll notice that it contains various pieces of information about the feature layer that was published, including the symbology and the data schema as it relates to the fields.

If you take a look at the URL for this REST service, it should look something like this:

```
http://services7.arcgis.com/LLWzNvydeNCpjeTo/arcgis/rest/services/BusSt
ops/FeatureServer/0
```

You'll notice in the path that you can see `rest`; this is the location of the services directory on your instance of ArcGIS Online. Say, you modify the URL to only show up until the `services` in the path, as follows:

```
http://services7.arcgis.com/LLWzNvydeNCpjeTo/arcgis/rest/services/
```

You will then be able to see the root directory for services on your ArcGIS Online account. From this base URL, you can navigate to all the services you have hosted on your ArcGIS Online account. If you were trying to access these services on ArcGIS Server, the (standard) URL for those services would be as follows:

```
http://<server host name>/arcgis/rest/services
```

This URL will be the base we use in the following section when it comes to using URL parameters to query and return data in a specific format.

URL parameters

In order to access data and format it accordingly through the ArcGIS REST API, you will have to provide one string that contains the REST endpoint along with various parameters known as the URL parameters. URL parameters are exactly what they sound like: parameters of the data you want to be returned in a specific format accessed through the Web using the URL as the query syntax. All URL parameters will be in the following format:

```
http://<rest-
endpoint>/operation?<parameter1=value1>&<parameter2=value2>
```

In the preceding example, `http://<rest-endpoint>` is the base URL for the feature or map service you are accessing on your ArcGIS Online or ArcGIS Server instance. The `operation` is the type of request you are making. For example, you would pass `query` if you would like to query the data. The `?` query string is used to indicate the beginning of the parameter list and `parameter1=value1` is called the name-value pair used to access data. In order to separate each name-value pair, you pass `&` in between each one.

Once we pass the request using specific parameters, the data will be returned in the browser in the default HTML format. For example, say, we query the Bus Stops feature where the Object ID is greater than 1:

```
http://services7.arcgis.com/LLWzNvydeNCpjeTo/arcgis/rest/services/BusSt
ops/FeatureServer/0/query?where=OBJECTID>1
```

Scroll down the web page and you should see the data returned as follows:

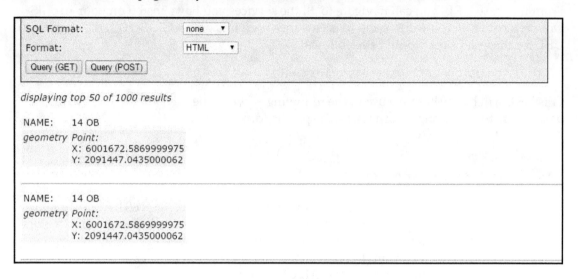

To learn about all the parameters you can pass to the URL query operation, look up Query–Feature Service (Operation) in the ArcGIS REST API reference.

If you pass a URL parameter without assigning the format to return the data, it will be returned in the HTML format by default. When requesting data from the ArcGIS REST API programmatically, we will want the data to be returned in various formats depending on our workflow. In order to return the data in a specific format, we will have to pass a parameter at the end of the URL. The ArcGIS REST API supports the response of URL parameters in various formats using the f format query parameter. For example, using the preceding URL request, we can get the data back in a format known as JSON or JavaScript Object Notation by passing f=json:

```
http://services7.arcgis.com/LLWzNvydeNCpjeTo/arcgis/rest/services/BusSt
ops/FeatureServer/0/query?where=OBJECTID>1&f=json
```

When you plug this into the browser, you should be able to see the data returned in JSON format as follows:

```
{"objectIdFieldName":"OBJECTID","globalIdFieldName":"","geometryF
{"wkid":102643,"latestWkid":2227},"fields":
[{"name":"NAME","type":"esriFieldTypeString","alias":"NAME","sqlT
OB"},"geometry":{"x":5999783.7865000069,"y":2088532.5639999956}},
OB"},"geometry":{"x":5999783.7865000069,"y":2088532.5639999956}},
OB"},"geometry":{"x":5999783.7865000069,"y":2088532.5639999956}},
OB"},"geometry":{"x":5999200.0789999962,"y":2087870.925999999}},{
OB"},"geometry":{"x":5999200.0789999962,"y":2087870.925999999}},{
OB"},"geometry":{"x":5999200.0789999962,"y":2087870.925999999}},{
IB"},"geometry":{"x":5997602.3225000054,"y":2086727.8235000074}},
IB"},"geometry":{"x":5997602.3225000054,"y":2086727.8235000074}},
IB"},"geometry":{"x":5997602.3225000054,"y":2086727.8235000074}},
IB"},"geometry":{"x":5998072.5035,"y":2086936.5470000058}},{"attr
{"x":5998072.5035,"y":2086936.5470000058}},{"attributes":{"NAME":
{"x":5999159.9105000049,"y":2087747.9344999939}},{"attributes":{"
{"x":5999159.9105000049,"y":2087747.9344999939}},{"attributes":{"
{"x":5999159.9105000049,"y":2087747.9344999939}},{"attributes":{"
{"x":5999785.0160000026,"y":2088452.4155000001}},{"attributes":{"
{"x":5999785.0160000026,"y":2088452.4155000001}},{"attributes":{"
{"x":5999785.0160000026,"y":2088452.4155000001}},{"attributes":{"
{"x":6001864.8349999934,"y":2091650.6574999988}},{"attributes":{"
```

JSON is a lightweight data exchange format usually used with JavaScript as the data format returned from an API. The ArcGIS REST API is also capable of returning other data formats, which include the following:

1. The ArcGIS JavaScript API, `f=jsapi`, does not work on feature services. You can view an example using the ESRI default service at `http://services.arcgisonline.com/arcgis/r est/services/world_physical_map/mapserver?f=jsapi`.

2. A layer package file of the published service, `f=lyr`, does not work on feature services, for example, `http://services.arcgisonline.com/arcgis/rest/services/world_physic al_map/mapserver?f=lyr`.

3. Help documents on the published service, `f=help`, does not work on feature services, for example, `http://services.arcgisonline.com/arcgis/rest/services/world_physical _map/mapserver?f=help`.

4. To return the JSON of your request in a format that is easily readable with spaces and line breaks, you can pass `f=pjson`, for example, `http://services.arcgisonline.com/arcgis /rest/services/world_physical_map/mapserver?f=pjson`.

Feature sets

A feature set is a lightweight data format that contains the feature schema (such as fields, the spatial reference, and the geometry type) and the actual data itself, much like a feature class. The feature set is a format that makes it easy and efficient to interact with GIS web services published on a server, including ArcGIS Online, Portal for ArcGIS, and ArcGIS for Server. If you run a tool or geoprocessing service using a server environment, the data format used in these processes is called feature sets and record sets. For the purposes of this chapter, we will be reviewing only feature sets.

It is important to note that when loading data into a feature set from a feature class, any updates on the feature set will affect the original feature class. For example, if you write a script that includes an update cursor or field calculation on the feature set, you will inadvertently affect the original feature class. We will review this, but the best practice to load data into a feature set is to create a feature class in-memory and then use that in-memory feature class to load data into the feature set. Let's begin by creating a `FeatureSet` in your script:

```
import arcpy
#create an empty feature set
f_set = arcpy.FeatureSet()
```

Once the `FeatureSet` object has been created, you can pass in a feature class to populate the feature set, for example, our `Bus_Stops` feature class:

```
import arcpy
#create a feature set using a feature class
f_set =
 arcpy.FeatureSet(r"C:\PythonBook\Scripts\SanFrancisco.gdb\SanFrancisco
\Bus_Stops")
```

Next, in order to see the data that is contained within the `FeatureSet,` we will run a search cursor as a dictionary comprehension on the data. Let's say we want to see the Bus Route name and Signage on the Bus for that route:

```
import arcpy
from pprint import pprint
#create a feature set using a feature class
f_set =
 arcpy.FeatureSet(r"C:\PythonBook\Scripts\SanFrancisco.gdb\SanFrancisco
\Bus_Stops")
bus_signage = {row[0]:row[1] for row in arcpy.da.SearchCursor(f_set,
  ["NAME","BUS_SIGNAG"])}
pprint(bus_signage)
```

In this script, I imported a Python library called `pprint`. This is essentially a pretty printer. It makes the output print in a nice readable format to your working console.

When you run the preceding script, you should see output that looks similar to this:

```
Select Windows PowerShell
PS C:\PythonBook\Scripts\Ch9> python featureset.py
{u'1 IB': u'Howard & Main',
 u'1 OB': u'Geary & 33rd Av.',
 u'10 IB': u'10 TOWNSEND',
 u'10 N': u'Sausalito Ferry',
 u'10 OB': u'De Haro & 17th St.',
 u'10 S': u'SFCC via Geary Blvd',
 u'108 IB': u'108 TREASURE ISLAND',
 u'108 OB': u'108 TREASURE ISLAND',
 u'12 IB': u'Van Ness Av.',
 u'12 OB': u'Chavez & Mission',
 u'120 S': u'COLMA BART',
 u'121 S': u'SKYLINE COLLEGE',
 u'122 N': u'STONESTOWN',
 u'122 S': u'SOUTH SF BART',
 u'130 S': u'AIRPORT BLVD&LINDEN',
 u'14 IB': u'Ferry Plaza',
 u'14 OB': u'Lowell St.',
 u'14L IB': u'Ferry Plaza',
 u'14L OB': u'San Jose Av., Daly City',
 u'14X IB': u'Ferry Plaza',
 u'14X OB': u'San Jose Av., Daly City'.
```

In this example, you can see how we are able to take a feature class, load it into a feature set, and then loop over each record to create a dictionary of key-value pairs containing the Bus Route and Bus Stop Signage. For example, the Bus Route `14L IB` (Inbound) will have a sign with the destination `Ferry Plaza`.

Feature set methods

There are two methods that are included in the arcpy feature set class: load and save. The `load` method allows you to load data into a feature set from either a feature class or a web service. Then, the `save` method allows you to save any data contained within a feature set to a feature class or a shape file. We will create a script that will take a feature service that is published on ArcGIS online, load it into a feature set, and then finally save that feature set as a feature class in a geodatabase.

Let's start the script with the necessary parameters we'll need:

 Make sure that you share your ArcGIS Online Feature Service with "Everyone" so you can access it from a script. To share, go to Item Details -> Share -> Everyone. Accessing a password-protected feature service will be covered later in this chapter.

```
import arcpy
#allow data to overwrite
arcpy.env.overwriteOutput = True
#ArcGIS Online Feature Service of SF Bus Stops
bus_stops_url="http://services7.arcgis.com/LLWzNvydeNCpjeTo/arcgis/rest
/services/BusStops/FeatureServer/0/query?
  where=OBJECTID>0&outFields=*&f=json"
#Create the feature set
bus_featureset = arcpy.FeatureSet()
#load the json data into the feature set
bus_featureset.load(bus_stops_url)
#Next we'll save the feature set data to a feature class in our
  geodatabase
bus_featureset.save(r"C:\PythonBook\Scripts\SanFrancisco.gdb\Chapter9Re
  sults\agol_bus")
```

Note how I pass the `outFields=*` parameter to the URL; if you don't do this, you won't get all the fields returned in your request. Once you run this script, you should see a feature class called `agol_bus` in your file geodatabase. If you view the contents pane of the feature class in ArcCatalog and click on preview and look at the data as a table, you can see that there are only 1000 records in our feature class. This is because on ArcGIS Server and ArcGIS Online the default max record count is set to 1000. So, when we pass the URL request to get data back, we can get up to 1000 records. There are two workarounds to deal with this issue: the first way is to create a loop that makes a request for each sequential 1000 features and append each request to your feature class; I will be using this method later in the chapter and the other option is to update the definition of the feature service. I don't always recommend that you do this because you can corrupt the definition of your feature service and it also does not always appear to work on every instance of ArcGIS Online. To give it a try, you can navigate to the `admin` directory of your services by replacing `rest` with `admin` and the "/" between your service name and `FeatureServer` with ".", like this: `http://services7.arcgis.com/LLWzNvydeNCpjeTo/arcgis/admin/services/BusStops.FeatureServer/0/`

Once you have navigated here, scroll to the bottom of the page and click on
UpdateDefinition. The next screen should appear as follows:

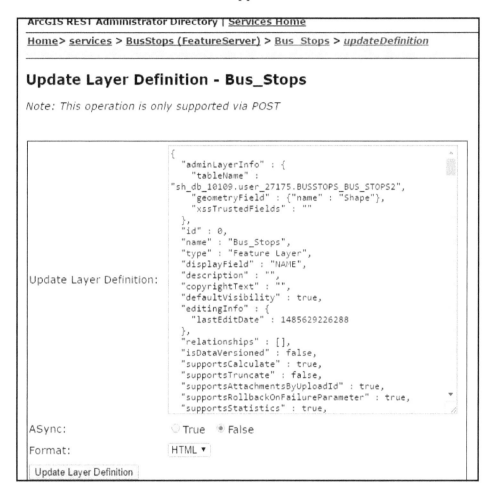

Under `lastEditDate`, change the value to `null`:

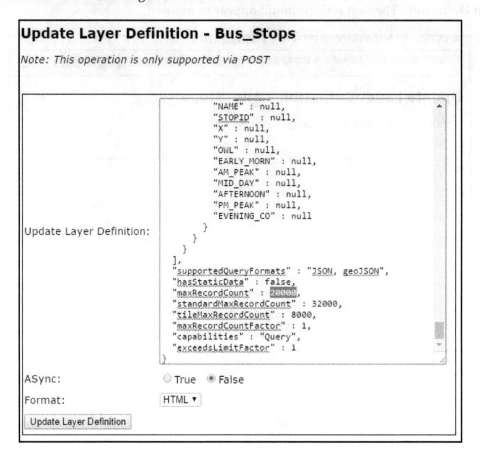

Then, scroll to the bottom and change the **MaxRecordCount** to your preferred value; in this case, I'll make it **20000**:

Finally, click on **Update Layer Definition** at the bottom of your screen. This does not always work, which is why I recommend that you use a loop to request data from the ArcGIS REST API. It is more successful on paid versions of ArcGIS Online than on any free or trial versions. In the next chapter, we will explore a third option to work with this limitation using the ArcREST Python package.

ArcGIS Online tokens

A token is a unique identifier that allows an application or user to access some part of an API. It is basically a complex string that is generated and provided to an application accessing the services provided by the API. In our case, essentially, a token allows an authenticated user access to private or password protected REST services through the ArcGIS REST API.

In order to see the output of an ArcGIS REST API token, we can see the response by passing the parameters it takes through an URL. Navigate to the browser and enter this URL: `https://www.arcgis.com/sharing/generateToken?username=youruser&password=yourpassword&referer=https://www.arcgis.com&f=json`

Replace `youruser` and `yourpassword` with your actual username and password for the ArcGIS Online account you have or set up earlier in the book. When you enter this URL, because we passed `f=json`, you should see something like this in the browser:

{"token":"KKKwZ5WTv57R7nrriRY243ABPXFk4KjIl0NrKE9a4mkeQG8PLSO73WzRnK MutIUzItaEJpYW47YGNWhUGkpCk7wDfqUTh8c_SH95CbB_0P-WA_vkjGx6tPDwCP7Y-MB4IH39yBRfxzYWIPohfPSc_A..","expires":1485668316048,"ssl":false}

This is what the output of an ArcGIS API token request looks like in a JSON format. Next, we will go over a Python script and function that you will be able to use to access protected services on your ArcGIS Online account.

The following Python code will demonstrate how to access a token to use an ArcGIS Online web service that is password-protected, because if the hosted feature services has not been shared with "Everyone", you will need to use a token to access it programmatically. The first thing we'll have to do to begin with is import the libraries we'll need to be working with in our `arcpy_token.py` script:

```
import urllib
import urllib2
import httplib
import json
import contextlib
```

You will see how each of these is used in the script as we begin to build it. Next, we will have to pass a username and password as parameters in the request. Because sometimes we might be sharing our script with other users or hosting it on a git version controlled site, we never want to put our username and password out there for someone to see. I like to store this sensitive information in separate files that can be accessed and assigned to variables in our script. To do this in Python, we can do this:

```
user_file = open('C:\coding\Python\GIS\username.txt', 'r')
username = user_file.readline().rstrip('\n')
pass_file = open('C:\coding\Python\GIS\password.txt', 'r')
password = pass_file.readline().rstrip('\n')
service_url = "https://arcgis.com/sharing"
```

In this example, we used the open function that allows us to read the contents of `username.txt`. Then, we can pass those contents to `username` and strip any return characters `'\n'` that may exist within the text file. We'll do the same for `password.txt`. It's important to note that you can store these files anywhere on your local machine and name them what you want in order to conceal this information from other users (note that this is not secure; refer to password security experts about how to best save your passwords). The `service_url` variable is used in the token function that we will discuss next.

Next, we'll create two functions that will provide us with the token. The first function is designed to process the URL request:

```
def submit_request(request):
    """ Returns the response from an HTTP request in json format."""
    with contextlib.closing(urllib2.urlopen(request)) as response:
        job_info = json.load(response)
        return job_info
```

The preceding function will be used in the larger function called `return_token`, as follows:

```
def return_token(service_url, username, password):
    """ Returns an authentication token for use in ArcGIS Online."""
    # Set the username and password parameters before
    #  getting the token.
    params = {"username": username,
              "password": password,
              "referer": "https://www.arcgis.com",
              "f": "json"}
    #build url for the generate token
    service_url = "{}/generateToken".format(service_url)
    #this is variable to be passed to the "submit_request" function
    request = urllib2.Request(service_url, urllib.urlencode(params))
    print "REQUEST IS ", request
    token_response = submit_request(request)
```

```
      print "TOKEN RESPONSE ", token_response
      #if condition to test if token is returned
      if "token" in token_response:
          print("Getting token...")
          token = token_response.get("token")
          return token
      else:
          # Test the request on HTTPS if error returned
          # Request for token must be made through HTTPS.
          if "error" in token_response:
              error_mess = token_response.get("error", {}).get("message")
              if "This request needs to be made over https." in
                error_mess:
                  token_url = token_url.replace("http://", "https://")
                  token = get_token(service_url, username, password)
                  return token
              else:
                  raise Exception("AGOL error: {} ".format(error_mess))
```

This function is what we will use to get a token to access a hosted feature service on ArcGIS online that has not been shared with "Everyone". The parameters that the function requires are `service_url`, `username`, and `password`. This function is basically building the URL like the one we used at the beginning of this section and returning the token so we can assign it to a variable as shown in the following steps:

1. First, we build the parameters we will be appending to our URL and assign them to the `params` variable.
2. Then, we append the `generateToken` operation to our service URL and reassign this value as our base service URL.
3. Next, we assign the format of our URL token request to the request variable. Then, we pass this `request` variable to the `submit_request` function we built earlier.
4. Finally, we provide an `if...else` condition to return the token if the request was successful, or try again using an `HTTPS` request, or provide the error that is returned in the request. The final token script called `arcpy_token.py` should look like this:

 In the code where you see `username.txt` and `password.txt`, make sure you have these text files that have user names and passwords to your ArcGIS Online account you created in the previous chapter.

```
import urllib
import urllib2
```

```
import httplib
import json
import contextlib
user_file = open('C:\coding\Python\GIS\username.txt', 'r')
username = user_file.readline().rstrip('\n')
pass_file = open('C:\coding\Python\GIS\password.txt', 'r')
password = pass_file.readline().rstrip('\n')
service_url = "https://arcgis.com/sharing"
def submit_request(request):
    """ Returns the response from an HTTP request in json format."""
    with contextlib.closing(urllib2.urlopen(request)) as response:
        job_info = json.load(response)
        return job_info
def return_token(service_url, username, password):
    """ Returns an authentication token for use in ArcGIS Online."""
    # Set the username and password parameters before
    #  getting the token.
    params = {"username": username,
              "password": password,
              "referer": "https://www.arcgis.com",
              "f": "json"}
    service_url = "{}/generateToken".format(service_url)
    request = urllib2.Request(service_url, urllib.urlencode(params))
    print "REQUEST IS ", request
    token_response = submit_request(request)
    print "TOKEN RESPONSE ", token_response
    if "token" in token_response:
        print("Getting token...")
        token = token_response.get("token")
        return token
    else:
        # Request for token must be made through HTTPS.
        if "error" in token_response:
            error_mess = token_response.get("error", {}).get("message")
            if "This request needs to be made over https." in
              error_mess:
                token_url = token_url.replace("http://", "https://")
                token = get_token(service_url, username, password)
                return token
            else:
                raise Exception("AGOL error: {} ".format(error_mess))
agol_token = get_token(service_url, username, password)
print agol_token
```

These two functions will be used in our larger script that we will begin reviewing and building in the next section.

Putting it all together

Everything that we have gone over in this chapter was to build the foundation of knowledge required to use Python and ArcPy to access ArcGIS Online web services. We reviewed the ArcGIS Online REST API, URL parameters, feature sets, and ArcGIS Online tokens. Now we will be using all these components to build a script that will take the location of points from ArcGIS Online and update the location of those points in a file geodatabase. Let's say you are a water utility and you have staff in the field updating the location and/or attribute information of your fire hydrant data. This is a useful way to automatically update your on-premise data with the data collected in the field. A lot of utility organizations consider their data sensitive and unless you have an ArcGIS Server setup in a DMZ environment, you may have to use ArcGIS Collector with ArcGIS Online for your field staff.

To begin this script, we will be importing both the arcpy library and the Python script we just created to access a token. Make sure that the `arcpy_token.py` script only has the import statements and is saved in the same directory as our final script:

```
import arcpy
#import the two functions from the arcpy_token.py script
from arcpy_token import return_token, submit_request
#make it possible to overwrite an existing feature class
arcpy.env.overwriteOutput = True
#assign the feature class we will be updating to the variable update
update=r"C:\PythonBook\Scripts\SanFrancisco.gdb\Chapter9Results\BusStop
  s_Moved_Update"
user_file = open('username.txt', 'r')
username = user_file.readline().rstrip('\n')
pass_file = open('password.txt', 'r')
password = pass_file.readline().rstrip('\n')
service_url = "https://arcgis.com/sharing"
```

The preceding code explanation is as follows:

1. To begin with, we assign the feature class that we will be updating, using the service from ArcGIS Online, to our `update` variable.
2. Then, we will be accessing our username and password that are saved in a text file and assigning them to our `username` and `password` variable.
3. Next, we will be assigning the base URL we use to access a token to the `service_url` variable.

4. Once complete, we will begin to build our main function and call it `update_featureclass_agol`.

5. This function is designed to loop over a request for data from an ArcGIS Online feature service and then load that data into a feature set, which will be appended to a feature class we will create in-memory. Using this `in_memory` feature class, we will update a feature class in a geodatabase using the x, y location from the ArcGIS Online feature service. Let's start building the function:

```
def update_featureclass_agol(base_URL, update_feature, count):
        n = 0
        template = r" C:\PythonBook\Scripts\SanFrancisco.gdb\
        SanFrancisco\Bus_Stops"
        FC = arcpy.CreateFeatureclass_management("in_memory",
        "FC", "POINT", template,"DISABLED", "DISABLED", "", "",
          "0", "0", "0")
        token = return_token(service_url, username, password)
```

Our function will be taking three parameters; the first one is called `base_URL`. The `base_URL` is the REST endpoint for our ArcGIS Online feature service that we will be using to append a query. The next parameter is the `update_feature`; this is the feature class we will be updating, which we assigned to the `update` variable earlier in the script. And the third parameter is called `count`; this will be the number of loops we will have to perform in order to access all the data in the feature service. If you recall from earlier, we can only access 1000 records per request, so we will need to loop the number of features. For example, if we have 5,000 records in our feature services, our `count` will be 5 (5 X 1000).

Our first local variable in the function is n with a value equal to 0. This variable will be used in the query statement for every request we make further on in the script. Next, we create a template feature class using our `Bus_Stops` and assign it to `template`. This will be used as the template for the schema in our `in_memory` feature class called `FC`. The `FC` variable will be where we run the create feature class tool; you'll notice that for the output parameter where we would usually assign the output name and location for our feature class, we use `in_memory`. The `in_memory` setting is a function of arcpy that allows you to run tools and output the data in memory without using any physical space on your computer hard drive. It is also much faster to run processes in memory rather than writing data to the hard drive. We do this because this is an intermediate dataset and we don't want to create unnecessary features on our hard drive to achieve the end goal. The `token` variable is assigned the output of our `return_token` function, which is the token we went over in the previous section.

In the next section of the function, we will create a loop that will query 1000 records per loop, load them into a feature set, and then append that feature set to our in-memory feature class:

```
#loop over request number of times where count is number of
    features/1000
for x in range(count):
    where = "OBJECTID>"+str(n)
    #url parameters
    query = "/query?where={}&returnGeometry=true&outSR=2227&
    outFields=*&f=json&token={}".format(where, token)
    fs_URL = base_URL + query #build the finla url
    fs = arcpy.FeatureSet()
    fs.load(fs_URL)
    arcpy.Append_management (fs, FC, "NO_TEST")
    n+=1000 #add 1000 to n for next query
```

This for loop will loop the number of times that we set the variable `count` as a parameter in our function. The `where` variable is the where statement that takes the `OBJECTID` value that is greater than the value of n we pass. For example, in the first loop, the where statement will be `OBJECTID > 0`, then for the second loop, the statement will be `OBJECTID > 1000`, and so forth. If you are managing a dataset where the `OBJECTID` skips or is not sequential, then you can use the `resultOffset` URL parameter to use pagination on your requests. So, say you're passing something like this:

```
'&resultOffset='+str(n)
```

You will be able to make requests that are in the order of 1,000, regardless of the `OBJECTID` order.

The next variable, `query`, is where we build the URL parameters for the query of the ArcGIS Online feature service. The `fs_URL` variable is where we concatenate these variables to construct our final URL for the ArcGIS REST API. In order to access the feature service, we'll have to load it into a feature set that is created and assigned to the `fs` variable. Then, we use the load method we reviewed earlier in the chapter to load the data from the ArcGIS API by passing `fs_URL` to the feature set. Finally, in this loop, we append each request that has been loaded into a feature set to our `in-memory` feature class `FC`. Once we have finished building our feature class from the ArcGIS Online service, we will use our `in-memory` feature class to update the x, y location of the feature class in our file geodatabase:

```
with arcpy.da.SearchCursor(FC, ['OID@', 'SHAPE@XY', "FACILITYID"]) as
cursor:
        for row in cursor:
            objectid = row[0]
            #get the x value from 'SHAPE@XY'
```

```
pointx = row[1][0]
#get the y value from 'SHAPE@XY'
pointy = row[1][1]
fid = row[2]
fid_sql ="FACILITYID = {0}".format(fid)
with arcpy.da.UpdateCursor(update_feature, ['SHAPE@'],
  fid_sql) as
  cursor:
    for urow in cursor:
        print "FACILITYID updated is ", fid
        #update the location of point using the x and y
          values from the FC feature
        urow[0] = arcpy.Point(pointx, pointy)
        cursor.updateRow(urow)
```

Finally, in our script, we'll write the main function that will call the function we wrote earlier:

```
def main():
update_featureclass_agol("http://services7.arcgis.com/LLWzNvydeNCpjeTo/
arcgis/rest/services/BusStops/FeatureServer/0", update, 17)
if __name__ == '__main__':
    main()
```

The final script should look like this:

```
import arcpy
from arcpy_token import return_token, submit_request
#make it possible to overwrite an existing feature class
arcpy.env.overwriteOutput = True
#assign the feature class we will be updating to the variable update
update =
 r"C:\PythonBook\Scripts\SanFrancisco.gdb\Chapter9Results\BusStops_Move
d_Update"
user_file = open('username.txt', 'r')
username = user_file.readline().rstrip('\n')
pass_file = open('password.txt', 'r')
password = pass_file.readline().rstrip('\n')
service_url = "https://arcgis.com/sharing"
def update_featureclass_agol(base_URL, update_feature, count):
    #set the variable n to 0 to be used to query objectID
    n = 0
    #template feature class to be used when creating feature class FC
    template =
      r"C:\PythonBook\Scripts\SanFrancisco.gdb\SanFrancisco\Bus_Stops"
    #create feature class in memory
    FC = arcpy.CreateFeatureclass_management("in_memory", "FC",
      "POINT", template, "DISABLED", "DISABLED", "", "", "0", "0", "0")
```

```python
        #generate token
        token = return_token(service_url, username, password)
        #loop over request number of times where count is number of
          features/1000
          for x in range(count):
            where = "OBJECTID>"+str(n)
            #url parameters
            query = "/query?where=
              {}&returnGeometry=true&outSR=2227&outFields=*&f=json&token=
              {}".format(where, token)
            fs_URL = base_URL + query #build the finla url
            fs = arcpy.FeatureSet()
            fs.load(fs_URL)
            arcpy.Append_management (fs, FC, "NO_TEST")
            n+=1000 #add 1000 to n for next query
            print n
    with arcpy.da.SearchCursor(FC, ['OID@', 'SHAPE@XY', "FACILITYID"])
      as cursor:
        for row in cursor:
            objectid = row[0]
            #get the x value from 'SHAPE@XY'
            pointx = row[1][0]
            #get the y value from 'SHAPE@XY'
            pointy = row[1][1]
            fid = row[2]
            fid_sql ="FACILITYID = {0}".format(fid)
            with arcpy.da.UpdateCursor(update_feature, ['SHAPE@'],
              fid_sql) as cursor:
                for urow in cursor:
                    print "FACILITYID updated is ", fid
                    #update the location of point using the x and y
                      values from the FC feature
                    urow[0] = arcpy.Point(pointx, pointy)
                    cursor.updateRow(urow)
def main():
  update_featureclass_agol("http://services7.arcgis.com/LLWzNvydeNCpjeTo/
arcgis/rest/services/BusStops/FeatureServer/0", update, 17)
if __name__ == '__main__':
    main()
```

Before we run the script, you should notice the difference in location between the
Bus_Stops feature class and the BusStops_Moved_Update feature class. They should look
like this:

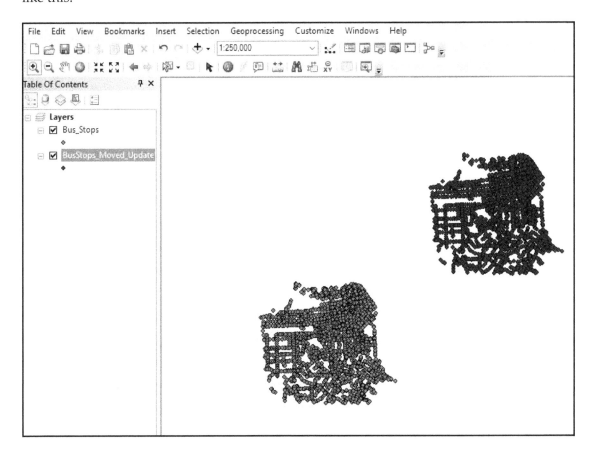

Close any applications that are accessing the `BusStops_Moved_Update` feature class and then run `final_script.py`. Once the script has completed, your `update` feature class should appear in the same location as the `Bus_Stops` feature class:

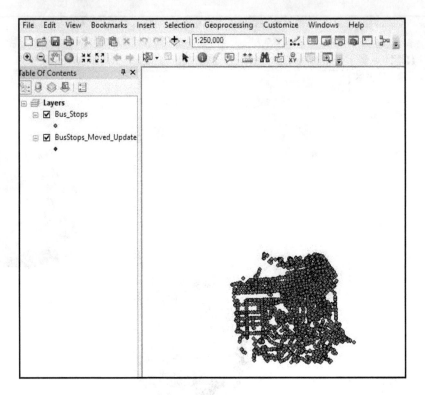

Summary

In this chapter, we covered the use of the ArcGIS REST API, feature sets, URL parameters, ArcGIS Online Tokens, and how to put them all together in one script. This workflow can be used in many examples within your organization. If you need to synchronize data from your ArcGIS Online account that is collected in the field with data in your office, you can use this process.

In the next chapter, we will cover the ArcREST Python package that allows us to execute Python with ArcGIS Online much more efficiently. This chapter was a good introduction to interacting with the ArcGIS REST API without any additional packages, but moving forward, we will begin introducing these new methods using the ArcREST Python package.

10
ArcREST Python Package

After the introduction to the ArcGIS REST API and beginning to work with Python and ArcGIS web services, we have now built the foundation of knowledge necessary to begin working with the ArcREST Python library developed by ESRI. This module is designed for working with the ArcGIS REST API and making it easier and more efficient. If you work with ArcGIS Online, Portal for ArcGIS, or ArcGIS Server, the ArcREST Python package will make managing services and the administration of your sites much easier. As you may remember from the previous chapter, we wrote a function called `return_token` that was approximately 14 lines of code in order to access a token for an ArcGIS feature service. Now, using ArcREST, we will show you how to get this token with only one line of code, along with many other use cases. Although you can use ArcREST with Portal for ArcGIS and ArcGIS for Server, in this chapter, we will primarily be covering use cases for ArcGIS Online.

This chapter will cover the following topics:

- Introducing the ArcREST module
- Installing ArcREST
- Introducing the ArcREST Helper module
- Administration using ArcREST
- Appending data to a hosted feature service
- Adding fields to a hosted feature service
- Updating the domain of a hosted feature service
- Querying a hosted feature service and saving as a feature class

Introducing the ArcREST module

The ArcREST package is a Python wrapper for the ArcGIS REST API. Before ArcREST, GIS Python developers working with the ArcGIS web infrastructure had to write multiple functions and many lines of code to administer the sites and web services related to that specific infrastructure. Now, with ArcREST, this has made developing with Python and the ArcGIS REST API much easier. You can find the ArcREST Python package, which was developed by ESRI, on GitHub at `https://github.com/Esri/ArcREST`.

 It is important to note that ESRI does not officially provide technical support for ArcREST.

Because ArcREST is not provided as a part of the typical ArcGIS installation like ArcPy, we will have to install this ourselves.

Installing ArcREST

There are two ways in which you can install the ArcREST Python package. The recommended method of installation is to use the `pip` method, which was covered earlier in the book. To install, we can just use the pip command `pip install arcrest_package`, as follows:

```
Windows PowerShell

Windows PowerShell
Copyright (C) 2016 Microsoft Corporation. All rights reserved.

PS C:\Users\Dara> pip install arcrest_package
```

After the installer has completed installation, your console should look similar to this:

If you work in an environment where it is difficult to use `pip` or you are not comfortable using `pip`, you can install ArcREST by downloading the source code from GitHub and modifying your system path in your scripts, as shown here:

```
import sys
sys.path.append(r"C:\GISProject") #path to the folder of ArcREST
```

The path in the preceding line of code should contain an ArcREST folder, so when you call ArcREST by typing `import arcrest`, it will allow you to access the functions available in this module.

Introduction to the ArcREST package structure

When installing ArcREST, it is important to know that there are actually two folders that need be installed in your working directory. Assuming you install this using `pip`, if you navigate to your `site-packages` directory in your Python installation `C:\Python27\ArcGIS10.5\Lib\site-packages`, you'll see two folders called `arcrest` and `arcresthelper`. These are the two directories you call when you use `import arcrest` or `import arcrest helper`. If you explore these folders, you can see all the functions and methods available through the modules. For example, if you navigate to `arcrest\security\security.py` and open that file, you can scroll down and see all the available methods in this part of the module.

In the `security.py` file, if you scroll down, you should see the class called
`AGOLTokenSecurityHandler`; you can read the comments about this function and see
what parameters are used as input:

The other folder we see included in the ArcREST package is called `arcresthelper`. This
folder contains modules and functions that are primarily used internally by other
operations within ArcREST. Two of the functions we will be using later in this chapter are
`securityhandlehelper.py` and `featureservicetools.py`.

ArcREST security handler

If you recall from Chapter 9, *ArcPy and ArcGIS Online*, we had to generate a token to access our password-protected hosted feature service. The function we wrote was called `return_token`, which returned a token that provided authentication for our script to access the web services on our ArcGIS Online account. In the ArcREST module, there is a built-in function called `AGOLTokenSecurityHandler`. This function enables us to generate a token by passing a few simple parameters. Depending on what piece of the ArcGIS platform you are developing, whether it is ArcGIS Online, ArcGIS Server, or Portal for ArcGIS, there are different parameters you will need to provide.

For ArcGIS Online security, take a look at the following code:

```
import arcrest
if __name__ == "__main__":
    username = raw_input("Enter user name: ")
    password = raw_input("Enter password: ")
    security_handler = arcrest.AGOLTokenSecurityHandler(login_info)
    print security_handler
```

As you can see in the preceding code, the security handler will take the username and password and provide the token that will be needed to access and administer your site. In the previous chapter, I provided an example where you could store your username and password in a separate file and access them using the `open()` and `read()` functions. Earlier, I provided another example of how you can pass your username and password. This uses the built-in Python function called `raw_input`, which allows you as a user to enter your parameters into the console instead of storing them somewhere on your computer hard drive. Other examples of using the security handler include use cases for ArcGIS Server and Portal for ArcGIS, as you can see here.

For ArcGIS Server security, take a look at the following code:

```
import arcrest
if __name__ == "__main__":
    token_url = "http://myserver_name:6080
    /arcgis/admin/generateToken"
    username = raw_input("Enter user name: ")
    password = "your_password"
    security_handler = arcrest.AGSTokenSecurityHandler(username,
      password, token_url)
```

When accessing Portal for ArcGIS security, the function is called
`PortalTokenSecurityHandler`. This function takes four parameters: username,
password, the token URL from your portal site, and the actual URL for your portal site, as
shown in the following code:

```
import arcrest
if __name__ == "__main__":
    username = raw_input("Enter user name: ")
    password = "your_password"
    token_URL =
        "https://myorgsite.com/portal/sharing/rest/generateToken"
    org_URL = "https://myorgsite.com/portal/sharing/rest"
    sh = arcrest.PortalTokenSecurityHandler(username,
                                            password,
                                            org_URL,
                                            token_URL)
```

As we can see from the preceding code, using the ArcREST security handler is much easier
and more efficient than building using the function we called `return_token.py` in the
previous chapter.

ArcGIS Online administration

If you are the administrator of your organization's ArcGIS Online account, ArcGIS Server,
or Portal for ArcGIS site, this next example is useful in accessing and administering
information for your site that can otherwise be redundant and tedious. As an example, let's
say you are an administrator of your organization's ArcGIS Online account and you want to
see the name, item ID, and sharing status of each of your services hosted on your account;
the following script will show you how to do that:

```
import arcrest
from arcresthelper import securityhandlerhelper
#build dictionary of log in info to be used in security handler
login_info = {'username': raw_input("Enter User Name: "), 'password':
raw_input("Enter Password: ")}
token = securityhandlerhelper.securityhandlerhelper(login_info)
#access admin rights
admin = arcrest.manageorg.Administration(securityHandler=
token.securityhandler)
content = admin.content
# Get the logged in user
user_info = content.users.user()
# List titles and sharing status for items in users' home folder
for item in user_info.items:
```

```
print item.id
print "[%s]\t%s" % (item.sharing['access'], item.title)
```

As you can see, we import both `arcrest` and `securityhandlerhelper` from `arcresthelper`. Then, we build a dictionary of the login info, our username, and the password. With that login info, we are able to access a token for the site. Next, we access the admin rights by passing our token to the `managerorg.Administration()` function in ArcREST. With the admin rights, we can get the content and the user info and print out the information that is contained within that user's home folder. This script is useful in a couple of ways: first, it might be helpful for you to know the sharing status of your services, and secondly, when using the ArcREST Python package, you will be required to pass the feature service **item ID** on multiple occasions. Having a list of all feature services and their associated item ID can be very useful. When we run the preceding script, you should see the following output in your console:

```
PS C:\pythonbook\scripts\ch10> python arcrest_started.py
Enter User Name: arcpybook
Enter Password:
92ff1991b38640aa84eaaa8bf94eaf0f
[private]        Census
938e672c08144ebfada1ab996719d4fb
[private]        Census
6bea8368781e49648c5638327ec9fb23
[private]        SF_Neignborhoods
eb03c4b243094edcb1ccc4b5f544173a
[public]         SF_Neignborhoods
91a321897de141b1b089aaf3fe7a75bc
[private]        BikeRoute
52a407c371fd4c34b9d90e27b521a6d7
[private]        BikeRoute
PS C:\pythonbook\scripts\ch10>
```

Querying hosted feature services

Now that we have covered how to access the functions of administration and security of our ArcGIS web infrastructure using ArcREST, we can introduce querying feature services. Although ArcREST is useful in administering your organization's site, it is also functional in querying and manipulating data that is already hosted in a production environment. In this section, we will demonstrate how to query and save all features to a feature class without having to loop like we did in the previous chapter. We will also cover how to append data to a feature service from a feature and change/update the schema of an existing feature service.

Querying all features and saving as a feature class

In the previous chapter, we had to design a script that would loop over a feature service because the record count was limited to 1,000. With the `QueryAllFeatures()` function available through ArcREST, you will no longer need to loop over the features unless you want to build the features in memory. Unfortunately, we cannot use the **in-memory** capability with the ArcREST module. The following script will show you how we can query a feature:

```python
import arcrest
from arcresthelper import featureservicetools
from arcresthelper import common
def main():
    #build a dictionary of the login info required for feature service
      tool
    login_info = {}
    login_info['security_type'] = 'Portal'#LDAP, NTLM, OAuth, Portal,
      PKI
    login_info['username'] = raw_input("Enter User Name: ")#<UserName>
    login_info['password'] = raw_input("Enter password: ")#<Password>
    login_info['org_url'] = "http://arcgis.com/"
    item_Id = "d4718e6a27a04deab6f764cd70e102f4"#<Item ID>
    sql = "OBJECTID>0"
    layerName = "Bus_Stops" #layer1, layer2
    saveLocation = r"C:\PythonBook\SanFrancisco.gdb\demo"
    #call the method to access feature service tools
    fea_service_tool =
      featureservicetools.featureservicetools(login_info)
    #using feature service access the individual feature service with
      the
    #Item ID
    feature_service = fea_service_tool.GetFeatureService(item_Id,False)
    print "Service is ", feature_service
    if feature_service != None:
        #Build the indivdual layer within a hosted feature service
        feature_service_url =
          fea_service_tool.GetLayerFromFeatureService
        (feature_service,layerName,True)
        print "url is ", feature_service_url
        if feature_service_url != None:
            #the main query function that will save the feature service
             as
            #Feature class
            demo =
              fea_service_tool.QueryAllFeatures(feature_service_url,
```

```
                                        where="1=1",
                                        out_fields="*",
                                        timeFilter=None,
                                        geometryFilter=None,
                                        returnFeatureClass=True,
                                        out_fc=saveLocation,
                                        outSR=None,
                                        chunksize=1000,
                                        printIndent="")
    if __name__ == "__main__":
        main()
```

In this script, we begin by importing our libraries and building our necessary login info to access the security handler. Then, we create the multiple variables that will be used in our final function called `QueryAllFeatures`. First, we define the SQL statement and pass it to the `sql` variable; this will be used to query the hosted feature layer; in this case, we want all the records so we'll pass `OBJECTID>0`. Next, we define the output location, including the feature class name, and pass it to the `saveLocation` variable. Once we've created the necessary parameters, we will begin to access the feature layer within the feature service.

In order to do this with ArcREST, we will first pass our login info to the `.featureservicetools()` function. Next, we'll use the `GetFeatureService()` method, followed by `GetLayerFromFeatureService()`. Finally, when we have our feature layer that we pass to the `feature_service_url` variable, we can run our `QueryAllFeatures()` function. This will automatically make it query in a loop until all records have been queried and then it will save the feature class to our `saveLocation` output.

This example demonstrates how we can use the ArcREST Python package to query a hosted feature layer without having to build a custom loop function like we did in the previous chapter. The main difference here, though, is that we can't access the `in_memory` capability we utilized previously.

Adding a field to a feature service

Sometimes, you might realize you need to add another field to a feature service you have hosted on ArcGIS Online. Traditionally, a user would have to download that feature service to a feature class, add the field, and then republish that feature service. With the ArcREST module, you can do this using one script without taking your data out of service. If you don't have it on your account, go ahead and publish the **BikeRoute** feature class to your ArcGIS Online account from **mxd**. Once we have the feature service published, we can build this script to add a field. Before we build this script, I want you to take a look at the attribute table as it currently exists before we run the script on the hosted feature service. If you go to your **My Content** section and click on the **BikeRoute** service you just published, you should see the item page with the feature description, and so on.

Click on the **Data** tab on the top of the page like this:

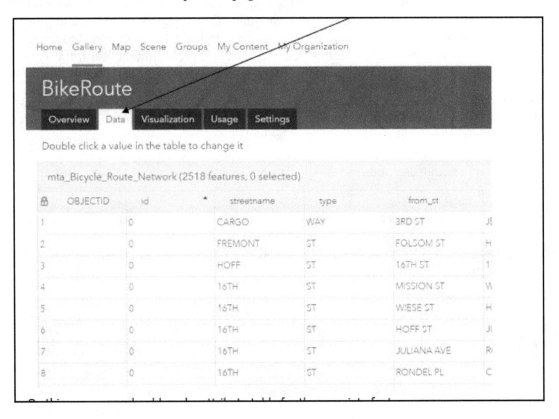

On this screen, you should see the attribute table for the associated feature service. Take some time to explore this page and scroll up/down and right/left to look at your data and fields. Now that you can see your schema, let's say we want to add a field called BIKEROUTE to our feature service without taking it down offline and changing the feature class and then republishing it. The following script will add that field for us:

```
from arcrest.security import AGOLTokenSecurityHandler
from arcrest.agol import FeatureLayer
if __name__ == "__main__":
    username = raw_input("Enter User Name: ")
    password = raw_input("Enter Password: ")
    #list of urls to add a field to
    #you can add this field to muiltple feature services if you'd like.
    #Just add them to this list
    urls = ["http://services7.arcgis.com/LLWzNvydeNCpjeTo/
    arcgis/rest/services/BikeRoute/FeatureServer/0"]
    #if you use a proxy generally for on site services like arcgis
      server
    #or portal
    proxy_port = None
    proxy_url = None
    #get the security handle token
    agol_security_handler = AGOLTokenSecurityHandler(username=username,
                                    password=password)
    #loop over the urls in the list of urls
    for url in urls:
        #create feature layer using feature layer
        feature_layer = FeatureLayer(
            url=url,
            securityHandler=agol_security_handler,
            proxy_port=proxy_port,
            proxy_url=proxy_url,
            initialize=True)
        #access admin rights to feature layer
        admin_feature_layer = feature_layer.administration
        field_to_add = {
            "fields" : [
                {
                    "name" : "BIKEROUTE",
                    "type" : "esriFieldTypeString",
                    "alias" : "Bike Route",
                    "sqlType" : "sqlTypeOther", "length" : 10,
                    "nullable" : True,
                    "editable" : True,
                    "domain" : None,
                    "defaultValue" : None
                } ]
```

```
    }
    #execute the add field method
    print admin_feature_layer.addToDefinition(field_to_add)
```

In this script, we can see some familiar operations as we import `arcrest` and use the ArcGIS Online security handler to get our token. Then, we pass the URLs to the feature services we want to add this field. Next, we will loop over each feature service in our list of services assigned to the `urls` variable. In each loop, we use the `FeatureLayer()` function to create our `feature_layer` object, and once that is created, we access the admin rights to that layer by using the `.administration` method. Once that is complete, we need to build the definition of our field and pass it to the `field_to_add` variable. Finally, we execute the final function called `.addToDefinition()`. This will take the field to add and append it to the `admin_feature_layer` we created.

Once you have executed your script, you should refresh your page, click on the **Data** tab again, and then look at the schema for your **BikeRoute** feature service. You should see the new **Bike Route** added to end of the table, as follows:

direct	cnn	number	Shape__Le...	Bike Route
2W	0	0	3,702.53835806205	
2W	0	0	632.440359104898	
2W	6,935,000	0	591.68093033827	
2W	726,000	0	235.835670386093	
2W	727,000	0	28.4436671436308	
2W	728,000	0	92.9587669668386	
2W	729,000	0	98.078594369969	
2W	730,000	0	41.1311860280713	
2W	731,000	0	146.389292812136	
2W	11,375,000	0	107.4171655694439	

Adding domains to fields in a hosted feature service

If you have altered the schema by adding fields, like we did in the previous section, you might want to add a domain to the feature service and assign it to your new field(s). The following script will demonstrate how we can add a domain to a field in an ArcGIS Online hosted feature service:

```
import arcrest
if __name__ == "__main__":
    #url for the admin site your accessing
    url =
      "https://services7.arcgis.com/LLWzNvydeNCpjeTo/arcgis/rest/admin"
    username = raw_input("Enter User Name: ")
    password = raw_input("Enter Password: ")
    #name of layer contained within feature service
    feature_layer_names = ["mta_bicycle_route_network"] # must be all
    #lowercase
    #definition of the domain
    definition = {
        "fields": [
            {
                "name": "BIKEROUTE",
                "domain": {
                    "type": "codedValue",
                    "name": "RouteType",
                    "codedValues": [
                        {
                            "name": "Option A",
                            "code": "type_a"
                        },
                        {
                            "name": "Option B",
                            "code": "type_b"
                        },
                        {
                            "name": "Option C",
                            "code": "type_c"
                        }
                    ]
                }
            }
        ]
    }
    #retrieve the security handle for the token
    security_handler = arcrest.AGOLTokenSecurityHandler(username, password)
```

```
#retrieve the services by passing url and token
agol_services = arcrest.hostedservice.Services(url, security_handler)
#loop over each service to find a field to assign domain to
for service in agol_services.services:
    print service
    #check if service has layers
    if service.layers is not None:
        print service.url
        #get the lyr name in service
        for lyr in service.layers:
            print lyr.name
            #if the layer is in your list as all lower update
            #the definition
            if lyr.name.lower() in feature_layer_names:
                print lyr.updateDefinition(definition)
```

In this script, the first thing we need to do is get the admin URL for your ArcGIS Online account; the URL is formatted as `https://(your agol server account)/arcgis/rest/admin`. Once we pass the admin URL and create variables for our password and username, we define the layer names in the service where we want the domain added. You can find the layer name under the service definition:

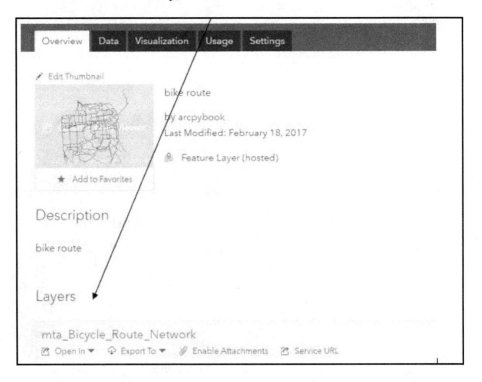

Next, we will define the domain that will be added to the service. This definition is in a **JavaScript Object Notation** (**JSON**) format because it will be added to the service definition. After defining the domain, we access the security handler, and with the token from the security handler we call the ArcGIS Online web service. If there are multiple layers in a service, we will loop over the service to find the layer name we defined earlier in the script. When we find the defined layer, we can run the `.updateDefinition()` method on the layer to add the domain definition we created earlier.

Once you have run the script, refresh the page, click on the **Data** tab, and scroll over to the **Bike Route** field we added in the previous section. If you click on a cell under the **Bike Route** field, you should see the options defined in our domain:

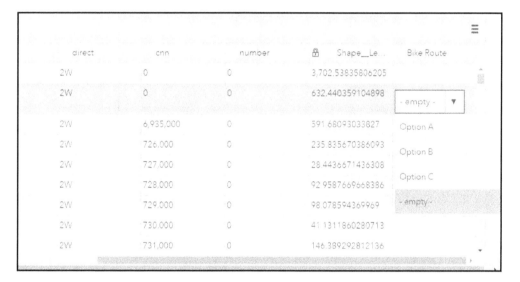

Adding a field and assigning a domain to that field is a common workflow that can be executed efficiently with the two previous scripts we covered here. This script can also be designed to search through all services within your organization's account, adding domains to the field in multiple layers.

Appending a feature class to a feature service

In many cases, we might want to append a feature service that is hosted in ArcGIS Online with data that has been added or updated on a feature class in a file or enterprise geodatabase. In the following script, we are going to append the bike route feature service with the exact same feature class; this will essentially double the feature count in our ArcGIS Online feature service. Before you run the following script, click on the **Data** tab and take a look at the total feature count:

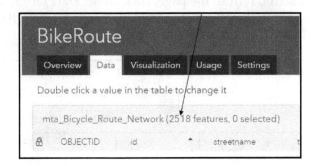

Now we can build the script and, following that, I will explain how it works:

```python
import arcrest, json
from arcresthelper import featureservicetool
from arcresthelper import common
if __name__ == "__main__":
    #create empty dictionary called "login_info"
    login_info = {}
    #create Key:Pair values for dictionary
    login_info['security_type'] = 'Portal'#LDAP, NTLM, OAuth, Portal, PKI
    login_info['username'] = raw_input("Enter User Name: ")#<UserName>
    login_info['password'] = raw_input("Enter Password: ")#<Password>
    login_info['org_url'] = "http://www.arcgis.com"
    item_Id = "52a407c371fd4c34b9d90e27b521a6d7"#<Item ID>
    layer_name='mta_Bicycle_Route_Network'#Name of layer in the service
    feature_class=r'C:\PythonBook\Scripts\SanFrancisco.gdb\SanFrancisco\
    mta_Bicycle_Route_Network'#Path to Feature Class
    atTable=None
    #activate the feature service tool
    fea_service_tool = featureservicetools.featureservicetools(login_info)
    if fea_service_tool.valid == False:
        print fea_service_tool.message
    else:
        feature_service =
            fea_service_tool.GetFeatureService(itemId=item_Id,
            returnURLOnly=False)
```

```
if feature_service is not None:
    feature_layer = fea_service_tool.GetLayerFromFeatureService
    (fs=feature_service,layerName=layer_name,returnURLOnly=False)
    if feature_layer is not None:
        #add the features from the feature class
        #to the feature service
        results = feature_layer.addFeatures
        (fc=feature_class,attachmentTable=atTable)
        print json.dumps(results)
    else:
        print "Layer %s was not found, please check
        your credentials and layer name" % layer_name
else:
    print "Feature Service with id %s was not found" % fsId
```

To start this script, we import all the necessary modules and then we build the login info necessary to access the token. Once that is done, we pass the item ID for the feature we are going to be appending to. To get the item ID, it is at the end of the URL when you click on the item or feature layer in AGOL. For example, when I click on the layer, I see this URL: `ht tp://arcpybook.maps.arcgis.com/home/item.html?id=0e8ea33c42f040aaa51ef845c0a 9cd23`; the long string following `id=` is the item ID.

Next, we pass the layer name to the `layer_name` variable and assign the feature class we will be using to append to the `feature_class` variable. Then, we begin to access the feature service tools provided through the ArcREST module. We first have to pass the login info to the feature service tools and then use the `GetFeatureService` method to access the feature service. Once we have the feature service, we will use `GetLayerFromFeatureService` to access the feature layer that is contained within the feature service. Finally, when we have the feature layer, we will use the `.addFeatures()` method to append all the features we have in our feature class table.

After you run the script, you should be able to refresh the page and see the count double for the features. This is because we are appending the same feature class as it exists in the hosted feature service for the purpose of this demo:

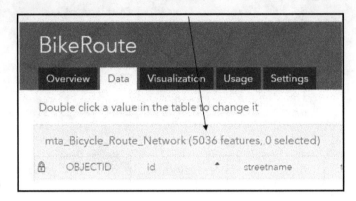

Updating records in a feature service

In this section, we will go over a script that will allow you as the user to update all records in a hosted feature layer table. For example, if you want to update all records in a table based on a SQL select statement, you can do that using the following script. In this example, we are going to update the street **type** for the **streetname** value of **SAN BRUNO**. Currently, in our data, all records for **SAN BRUNO** have a street **type** of **AVE**; we are going to write a script that will update all records of **AVE** that are associated with **SAN BRUNO** and change them to **WAY**.

The following screenshot shows how the table currently exists in our ArcGIS Online hosted feature service:

The script we use to complete this workflow is as follows:

```
from arcresthelper import securityhandlerhelper
from arcrest.agol import FeatureLayer
from arcrest.common.filters import LayerDefinitionFilter
import datetime
if __name__ == "__main__":
    # URL to Service
    url = 'http://services7.arcgis.com/LLWzNvydeNCpjeTo/
      arcgis/rest/services/BikeRoute/FeatureServer/0'
    sql = "streetname = 'SAN BRUNO'"# where clause
    fieldInfo =[
                {
                    'FieldName':'type',
                    'ValueToSet':'WAY'
                }
              ]
    securityinfo = {}
    securityinfo['security_type'] = 'Portal'#LDAP, NTLM,
    #OAuth, Portal, PKI, ArcGIS
    #User Name
    securityinfo['username'] = raw_input("Enter User Name: ")
    #password
```

```python
securityinfo['password'] = raw_input("Enter Password: ")
securityinfo['org_url'] = "http://www.arcgis.com"
sec_handle = securityhandlerhelper.securityhandlerhelper
(securityinfo=securityinfo)
if sec_handle.valid == False:
    print sec_handle.message
else:
    #create a feature layer of the AGOL service
    feature_layer = FeatureLayer(
        url=url,
        securityHandler=sec_handle.securityhandler,
        proxy_port=None,
        proxy_url=None,
        initialize=True)
    out_fields = ['objectid']
    #append the field info to each field
    for field in fieldInfo:
        out_fields.append(field['FieldName'])
    #query the feature layer
    query_feats = feature_layer.query(where=sql,
                        out_fields=",".join(out_fields))
    #loop over each feature and field to update
    for feat in query_feats:
        for field in fieldInfo:
            feat.set_value(field["FieldName"],field['ValueToSet'])
    #update features
    print feature_layer.updateFeature(features=query_feats)
```

After importing the libraries and creating variables for the security credentials, we provide the URL to the feature layer and build the SQL statement we want to use to update the records. Next, we create the `field_info` variable and assign the field and the value we want to set the field to in our update. In this example, we will select all records that have SAN BRUNO as `streetname` and update the `type` field with the WAY value. Then, we build our log info and provide it to the security handler in order to return our token. This part of the script should look familiar to you at this point in the chapter. We will then create an `if...else` condition to check whether the security handler is valid. Once we confirm that it returned as valid, we begin updating the records.

To do this, we will use the `FeatureLayer()` function to access the feature layer of the URL we pass. Then, we build the fields we want to update with the `outfields` variable. Next, we query the features, which will essentially select the records we want to update based on the SQL statement we defined earlier. Finally, we use the `.updateFeature()` method to update the records we selected using the `query_feats` variable.

After you have run the script, you should refresh the screen and see your records being updated similar to this:

Summary

In this chapter, we introduced the ArcREST Python package that can be used to efficiently script common tedious tasks in the ArcGIS web infrastructure, from ArcGIS Online to Portal for ArcGIS, and ArcGIS for Server. Although ArcREST is commonly used for administrative tasks in the ArcGIS web infrastructure, we demonstrated use cases that can be adapted by any GIS technician or analyst throughout an organization. We want to make sure you understand that scripting for ArcGIS Online or ArcGIS for Server does not have to be designated to administrators within an organization. Hopefully, the information and examples provided in this chapter provide you with enough foundational knowledge to apply ArcREST to multiple use cases within your organization. In the next chapter, we will begin to explore Python for ArcGIS Pro and introduce ESRI's migration of Python from the current Python 2.7 for ArcMap to Python 3.5 for ArcGIS Pro and the new Python API. This will be the beginning of the future for Python development in ESRI's ArcGIS platform.

11
ArcPy and ArcGIS Pro

After reviewing ArcPy and the ArcGIS Online infrastructure, we are shifting gears back to the latest desktop product release from ESRI, which is called ArcGIS Pro. ArcGIS Pro is the most recent release of ArcGIS desktop apps that speeds up processing and improves 3D and 2D **cartographic** representation. In this chapter, we will review some of the significant differences between ArcGIS Pro and ArcMap. The majority of the material covered in this chapter reviews the differences between using ArcPy in ArcGIS Pro as compared to the traditional ArcMap environment. We will discuss some of the significant architectural changes within the Python environment, as well as the use of a Python package manager called **Conda**.

In this chapter, we will cover the following topics:

- Introducing ArcGIS Pro
- ArcGIS Pro configuration
- Python 3.5 and ArcGIS Pro
- Python installation and configuration
- ArcGIS Pro Python window
- Conda

Introducing ArcGIS Pro

ArcGIS Pro was released in January 2015, and ESRI has called it the "Next Generation" of desktop applications for all GIS users. It is important to acknowledge that ArcGIS Pro is not considered a replacement for ArcMap but rather a powerful addition to the ESRI platform. ArcGIS Pro is designed to work seamlessly along with all other ESRI desktop apps, including ArcMap, ArcCatalog, and ArcScene. I would even say that ArcGIS Pro has fused some of the most used functionality of all three into one. In this chapter, we will go over some of the significant differences between ArcGIS Pro and the rest of the ESRI desktop app suite, especially as it relates to Python. For example, one significant change with the release of ArcGIS Pro was that ArcGIS Pro 1.4, the current version, relies on Python 3.5.2. Up until this point in the book, everything we have worked on regarding ArcPy and ArcMap has relied on Python 2.7.10. Although there are some changes between the two versions of Python as they relate to strings, dictionaries, and the print statement, most of the core concepts and syntaxes are the same between versions. We highlight some of the differences between Python 2 and 3 later in this chapter.

Installing and configuring ArcGIS Pro

ArcGIS Pro is dependent on ArcGIS Online, so in order to have ArcGIS Pro running on your desktop, you must have an ArcGIS Online account. You can either download a trial version of ArcGIS Pro or you can purchase the personal ESRI yearly license, which should also provide you with access to Pro. Regardless of how you intend to install and use ArcGIS Pro, the procedure should be standard across the board. As I said earlier, you must have an ArcGIS Online account. When you sign in to ArcGIS Online as the administrator, you should be able to click on **My Organization** and provide authorization for each named user to use ArcGIS Pro. If you are the administrator and you don't see **My Organization** in your ArcGIS Online account, you might need to contact ESRI to make sure it is set correctly for your license level.

Before working in the ArcGIS Pro environment, it is important to understand a few significant differences between ArcMap and ArcGIS Pro as they relate to GIS data.

In ArcGIS Pro, there are a number of data types that are not supported; these include personal geodatabases, geometric networks, raster catalogs, geodatabase servers, ArcMap document templates, ArcReader documents, tiled map packages, graphs, and topologies. When migrating scripts from ArcMap to ArcGIS Pro, it is important to remember that if any of these data types are in your scripts, they will not work properly.

When you first open ArcGIS Pro, you will notice a significant difference between ArcMap and ArcGIS Pro. To begin with, ArcGIS Pro works in **Projects**. A project functions like a working directory for all data related to your document. The project folder can contain maps, layout, tasks, database connections, toolboxes, and so on. The initial screen when you launch ArcGIS Pro looks like this:

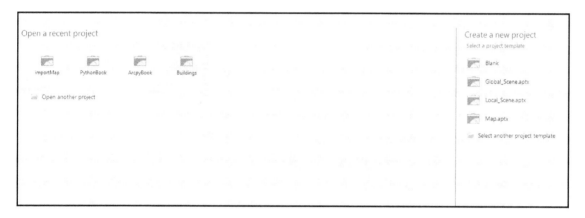

You can either open an existing project or create a new project. As you can see, there are four options to choose from under **Create a new project**:

- The first option is **Blank**, which will create a brand new project with no template associated with it.
- The second option is **Global_Scene.aptx**; this is a project template used to work with 3D data that expands a large geographic area.
- The third option is **Local_Scene.aptx**, the project template for 3D data that covers a smaller geographic region, for example, a city or county.
- Finally, you can see **Map.aptx**; this is the project template used to create 2D maps in ArcGIS Pro. The **.aptx** is the new file extension that is used for project templates in ArcGIS Pro.

If you choose `Map.aptx` as a template and then name your project, ArcGIS Pro should look similar to this:

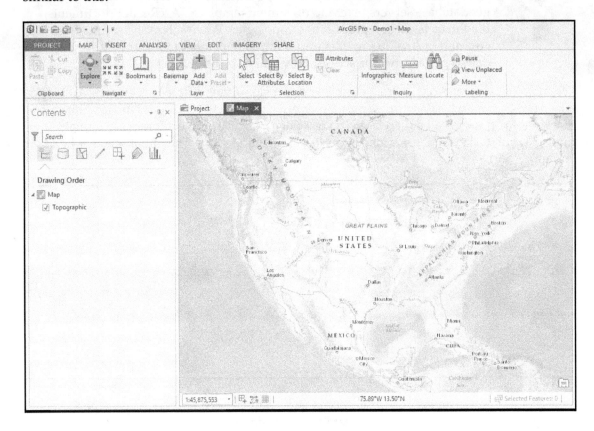

One of the first things you should notice that is different with ArcGIS Pro is the ribbon-based menu option at the top. We will not get into too many details regarding the ArcGIS Pro interface, but if you want to be able to use Python from ArcGIS Pro, you will need to open the Python window.

The ArcGIS Pro Python window

Click on the **Analysis** tab at the top of the ribbon and choose the **Python** option:

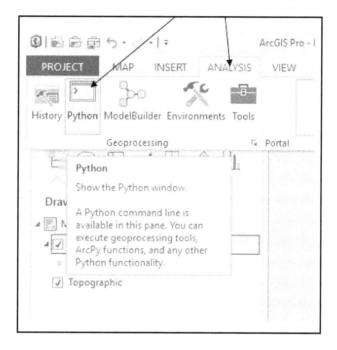

You should see the **Python** window appear at the bottom of the interface:

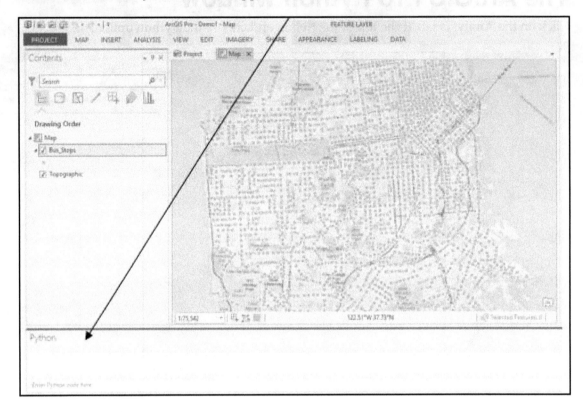

There are two regions of the Python window; they are called **Python Prompt** and **Transcript**. In the following figure, you can see each region if we pass a **print** function:

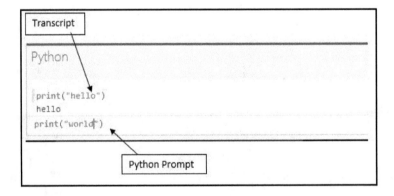

You should notice how the **print** function is slightly different here from its appearance in Python 2.7. In Python 3, instead of typing `print "hello world"`, you have to pass the string to a print function, for example, `print("hello world")`. This is one of the differences between Python 2 and Python 3, which we will cover in more detail in the following section.

Let's look at how else we can use the Python window in ArcGIS Pro. In your instance of Pro, add the **Bus_Stops** feature class from our working geodatabase. Then, in the Python window, type the following:

```
fields = arcpy.ListFields()
```

The following screenshot represents the previous code implemented in ArcGIS pro:

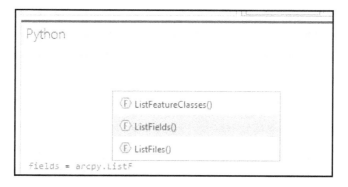

As you type in the prompt, you should begin to see the code completion for functions available in ArcPy. Next, instead of typing the name of the layer in the map document, you can actually drag it from the table of contents into the function parameter like this:

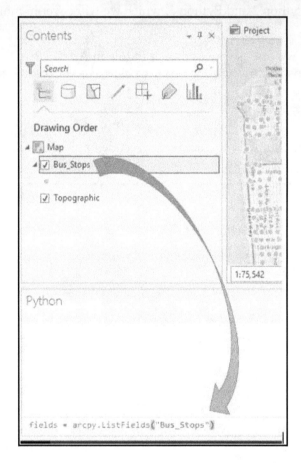

Next, hit *Enter* and then type in a loop to print each field name in the layer:

```
for field in fields:
print(field.name)
```

The following screenshot represents the previous code implemented in ArcGIS Pro:

```
Python

fields = arcpy.ListFields("Bus_Stops")
for field in fields:
    print(field.name)
```

Hit *Enter* twice and you should see all the field names print out. Here you can see the name for each field in the feature class printed:

```
Python

for field in fields:
    print(field.name)
OBJECTID
Shape
AGENCYID
AGENCY
ROUTEID
SEQUENCE
```

With this Python window, you can drag tools from the toolbox into the prompt and the snippet of code that this tool requires will appear as follows:

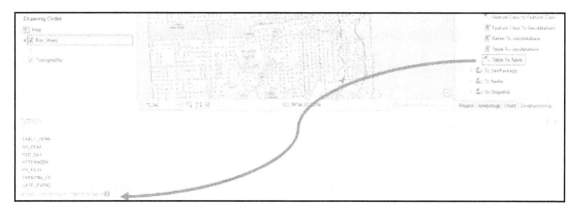

You can also drag files from your local computer to be used as parameters:

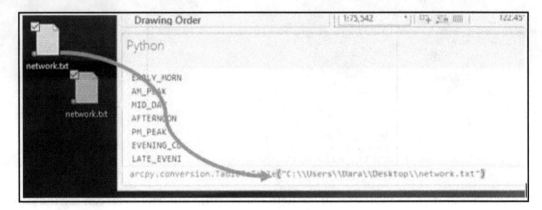

Python 2.7 and Python 3.5 with ArcPro

Everything we have covered so far in this book has been dependent on Python 2.7; however, both ArcGIS Pro and the new ArcGIS Python API have begun migrating to Python 3.5. If you are still using ArcMap, then you will continue to use Python 2.7 in that application. For GIS professionals and novices alike, you will have two versions of Python installed on your computer. The upcoming sections will explain how this is configured and how to work with these two environments separately.

During the installation of ArcGIS Pro, the software installation actually creates a directory and installs Python 3.5. To find this installation, you will have to navigate to `C:\Program Files\ArcGIS\Pro\bin\Python`.

In this directory, you should see `python.exe` and other folders, such as the `site-packages` folder. If you double-click on `python.exe` in this directory, you should see the Python command line appear like this:

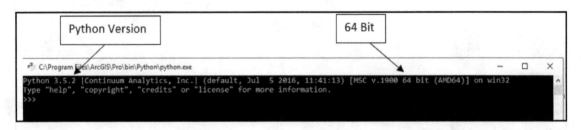

This is the location for the Python 3.5 installation; it is important to understand that this is *not* the actual executable used when you call Python from ArcGIS Pro. I will explain this in more detail in the following section, but the installed Python version will not actually work with ArcPy. For example, if we type `import arcpy` into this command prompt, we will get an error like this:

```
C:\Program Files\ArcGIS\Pro\bin\Python\python.exe                                          —
Python 3.5.2 |Continuum Analytics, Inc.| (default, Jul  5 2016, 11:41:13) [MSC v.1900 64 bit (AMD64)] on win32
Type "help", "copyright", "credits" or "license" for more information.
>>> import arcpy
Traceback (most recent call last):
  File "<stdin>", line 1, in <module>
ImportError: No module named 'arcpy'
>>>
```

Again, we will discuss in more detail how this version of Python is used in the new environment.

The 2.7 installation that we have been using for most of the book should be unaffected by this other version, and will be located at `C:\Python27\ArcGIS10.5`.

There are a few things you will need to know about the differences between Python 2.7 and Python 3.5. The first major change that will affect most scripts is the difference between the `print` statement and the `print` function. In Python 2, you would write a print statement like this:

```
print "Hello World"
```

This used to work in Python 2.7, but in Python 3, print is now a function and needs to be typed like this:

```
print("Hello World")
```

If you are working in both ArcGIS 10.5 and ArcGIS Pro, there are a couple of options that allow you to not rewrite all your code to work in both environments. The first option is to use the **2to3** Python tool that will update any Python script to be formatted for Python 3. The 2to3 module should exist in your current installed Python if you navigate to the following:

```
C:\Python27\ArcGIS10.4\ToolsScripts
```

You should see the `2to3.py` script located in there. To run this from the command line, type in the following:

```
PS C:\coding\python> python
C:\Python27\ArcGIS10.5\ToolsScripts2to3.py -w demo.py
```

This will update the demo.py file to be formatted for Python 3 and create a demo.bak backup file of your original script.

Another workflow you can implement to work with both environments uses the __future__ library:

```
PS C:\coding\python> python
Python 2.7.10 (default, May 23 2015, 09:40:32) [MSC v.1500 32 bit (Intel)] on win32
Type "help", "copyright", "credits" or "license" for more information.
>>> 8/7
1
>>> print "hello"
hello
>>>
>>> from __future__ import print_function
>>>
>>> print("hello")
hello
>>>
>>> from __future__ import division
>>> 8/7
1.1428571428571428
>>>
```

If you include the __future__ library in your 2.7 Python code, you can write the code so that it is compatible in both environments.

Conda and ArcGIS Pro

Beginning with ArcGIS Pro 1.3, ESRI introduced the implementation of Conda, a Python package manager. Conda is an open source package and environment management tool used to implement various versions of Python and dependencies. As you begin to work with Python on a regular basis, you will encounter situations where you need to have different versions of Python packages installed in safe locations that do not affect your main installation. Traditionally, Python developers have used something called "Virtual Environment" or "Virtualenv". Virtual environments create a separate installation of Python and associated packages in a designated location that you can call from the command line. It allows you, as a developer, to isolate different versions of Python and/or associated packages for specific projects. Over the last few years, Conda has gained popularity among Python developers as a package and environment manager.

When you run Python from ArcGIS Pro or as a standalone script, you are using a Python installation in a virtual environment created by Conda, called **arcgispro-py3**. If you remember, in the previous section, I explained that the Python executable located in C:\Program Files\ArcGIS\Pro\bin\Python was not used by the ArcGIS Pro Python window. It is this virtual environment called **arcgispro-py3** created by Conda that is used in ArcGIS Pro.

If you navigate to `C:\Program Files\ArcGIS\Pro\bin\Python\envs\arcgispro-py3`, you can see the Python directory used by ArcGIS Pro.

Running standalone scripts with Conda

In this section, I will discuss a couple of options available if you have to run standalone Python scripts from the Conda **arcgispro-py3** environment. Running a standalone script means running any script you have created is not able to go through the ArcGIS Pro Python window. First, it is important to know that you must run the Python command prompt as an administrator in order to take advantage of the Conda environment.

Click on `Start -> ArcGIS -> Python Command Prompt`:

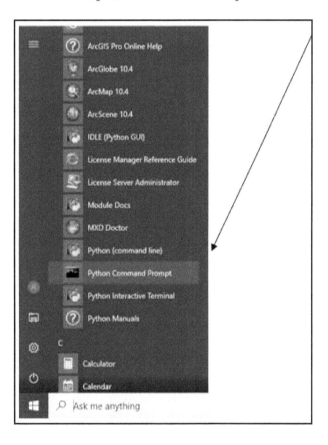

If you are not logged in as administrator, you will see this response in the terminal:

```
Access is denied.
Insufficient privileges to use Conda. Please open a Python Command Prompt
session with administrative privileges by right clicking on the link and
selecting "Run as Administrator".
```

Here is the actual representation of the terminal:

Note that the following part only applies to you if you are not an administrator on your machine. Before you log in or run as administrator from your user account, you will have to log in to your administrator account on your computer and sign into ArcGIS Pro in order to enable offline licensing. If you do not do this, when you try and run a script from the command line that has ArcPy in it, you will see this:

```
RuntimeError: Not signed into Portal
```

We can see the error in the command prompt:

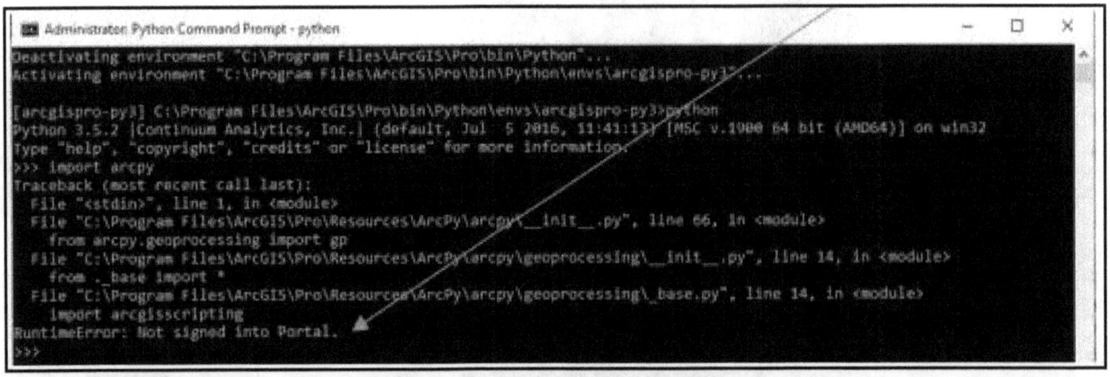

Next, we will go over how to enable offline licensing in your administrator account so you can run standalone scripts. Log in to your administrator account and launch ArcGIS Pro. Then, open a project template and click on the **Project** tab on the left-hand side:

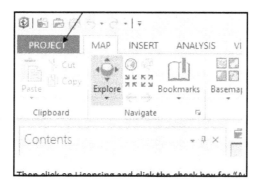

Then, click on **Licensing** and click on the checkbox for **Authorize ArcGIS Pro to work offline**:

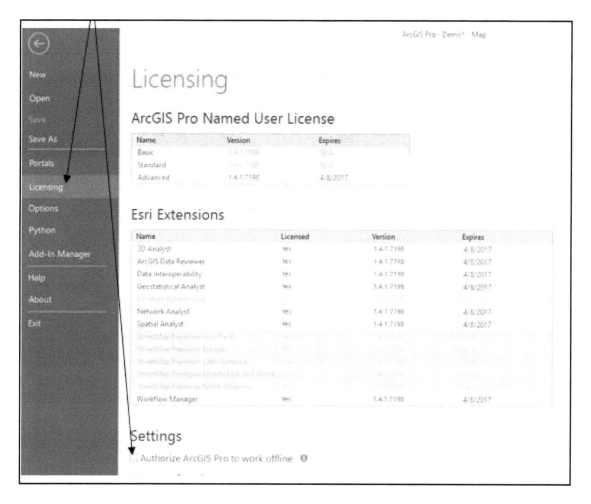

Now if you log back in to your account and run the Python command prompt as an administrator (by right-clicking on `run as administrator`), you should be able to run scripts that contain ArcPy. To test, open the Python command prompt as an administrator and then, at the prompt, type `python` to call `python.exe`:

```
Administrator: Python Command Prompt - python                                    —
Deactivating environment "C:\Program Files\ArcGIS\Pro\bin\Python"...
Activating environment "C:\Program Files\ArcGIS\Pro\bin\Python\envs\arcgispro-py3"...

[arcgispro-py3] C:\Program Files\ArcGIS\Pro\bin\Python\envs\arcgispro-py3>python
Python 3.5.2 |Continuum Analytics, Inc.| (default, Jul  5 2016, 11:41:13) [MSC v.1900 64 bit (AMD64)] on win32
Type "help", "copyright", "credits" or "license" for more information.
>>>
```

Then, type `import arcpy` and hit *Enter*. If you see a new chevron appear, like the one shown next, without an error, then that means it worked properly and you are ready to run standalone scripts using the arcgispro-py3 conda environment:

```
Administrator: Python Command Prompt - python                                    —
Deactivating environment "C:\Program Files\ArcGIS\Pro\bin\Python"...
Activating environment "C:\Program Files\ArcGIS\Pro\bin\Python\envs\arcgispro-py3"...

[arcgispro-py3] C:\Program Files\ArcGIS\Pro\bin\Python\envs\arcgispro-py3>python
Python 3.5.2 |Continuum Analytics, Inc.| (default, Jul  5 2016, 11:41:13) [MSC v.1900 64 bit (AMD64)] on win32
Type "help", "copyright", "credits" or "license" for more information.
>>> import arcpy
>>>
```

Another option to run standalone scripts is to use PowerShell as the administrator. Start PowerShell as the administrator and then run this command to run this `.bat` file:

```
&("C:\Program Files\ArcGIS\Pro\bin\Python\Scripts\proenv.bat")
```

The following screenshot represents the previous code in the PowerShell:

```
Administrator: Windows PowerShell
PS C:\WINDOWS\system32>
PS C:\WINDOWS\system32> &("c:\Program Files\ArcGIS\Pro\bin\Python\Scripts\proenv.bat")
Deactivating environment "C:\Program Files\ArcGIS\Pro\bin\Python"...
Activating environment "C:\Program Files\ArcGIS\Pro\bin\Python\envs\arcgispro-py3"...

[arcgispro-py3] C:\Program Files\ArcGIS\Pro\bin\Python\envs\arcgispro-py3>
```

We use the ampersand and parenthesis & ("..") because the file path has a space in it. You will notice that regardless of what directory you are in, when you start up the Conda environment, it will always put you in the C:\Program Files\ArcGIS\Pro\bin\Python\envs\arcgispro-py3 directory. You will just have to cd into another directory each time you activate the environment:

```
Administrator: Windows PowerShell

[arcgispro-py3] C:\Program Files\ArcGIS\Pro\bin\Python\envs\arcgispro-py3>cd C:\coding

[arcgispro-py3] C:\Coding>
```

I have a Python file called demo.py in C:\coding that just printed **hello world**:

```
print("hello world")
```

I can run it by typing python demo.py in the command line:

```
Administrator: Windows PowerShell

[arcgispro-py3] C:\Program Files\ArcGIS\Pro\bin\Python\envs\arcgispro-py3>cd C:\coding

[arcgispro-py3] C:\Coding>python demo.py
hello world

[arcgispro-py3] C:\Coding>
```

You will notice in the preceding image that arcgispro-py3 is at the beginning of the terminal prompt. This is the name of the virtual environment that we have activated. If you want to exit your Python virtual environment, you can just type exit():

```
[arcgispro-py3] C:\Program Files\ArcGIS\Pro\bin\Python\envs\arcgispro-py3>exit()
PS C:\WINDOWS\system32>
```

If you do not want to type in this long command each time to activate the Conda environment from PowerShell, you can create a batch file in your working directory to call the Conda environment. For example, in my working directory in `C:\coding\python`, I can create a file called `activateproenv.bat`, and in that file, I will type just one line:

```
call &("C:\Program Files\ArcGIS\Pro\bin\Python\Scripts\proenv.bat")
```

Then, I can just type in `.activateproenv.bat`, and it will activate the ArcGIS Pro Python Conda environment:

```
Administrator: Windows PowerShell - python                          —   □   ×
Windows PowerShell
Copyright (C) 2016 Microsoft Corporation. All rights reserved.

PS C:\WINDOWS\system32> cd C:\coding\python
PS C:\coding\python> .\activateproenv.bat

C:\coding\python>call    & ("c:\Program Files\ArcGIS\Pro\bin\Python\Scripts\proenv.bat")
Deactivating environment "c:\Program Files\ArcGIS\Pro\bin\Python"...
Activating environment "c:\Program Files\ArcGIS\Pro\bin\Python\envs\arcgispro-py3"...

[arcgispro-py3] C:\coding\python>python
Python 3.5.3 |Continuum Analytics, Inc.| (default, Feb 22 2017, 21:28:42) [MSC v.1900 64 bit (AMD64)] on win32
Type "help", "copyright", "credits" or "license" for more information.
>>> import arcpy
>>>
```

A third option to work with Conda and Python is to configure an IDE to work with your ArcGIS Pro Python environment. I would recommend that you use Spyder because it is very easy to install. Just make sure you are in your `arcgispro-py3` environment and pass the `conda install spyder` command:

```
PS C:\WINDOWS\system32> &("c:\Program Files\ArcGIS\Pro\bin\Python\Scripts\proenv.bat")
Deactivating environment "C:\Program Files\ArcGIS\Pro\bin\Python"...
Activating environment "C:\Program Files\ArcGIS\Pro\bin\Python\envs\arcgispro-py3"...

[arcgispro-py3] C:\Program Files\ArcGIS\Pro\bin\Python\envs\arcgispro-py3>conda install spyder
Fetching package metadata: ......
Solving package specifications: .........
```

When the installation has finished running, you can see where Spyder has been installed by typing `where spyder`; to actually launch Spyder using our virtual environment, just type `spyder`:

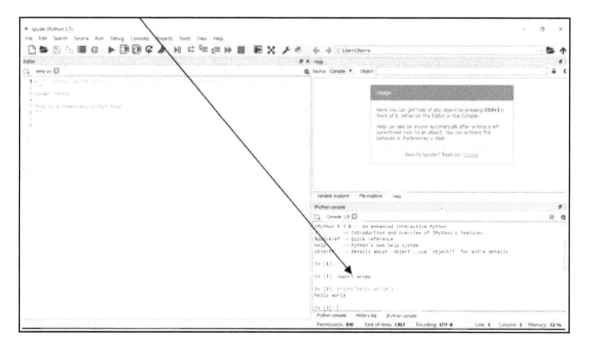

Once it launches, you should be able to type **import arcpy** in the console without receiving an error:

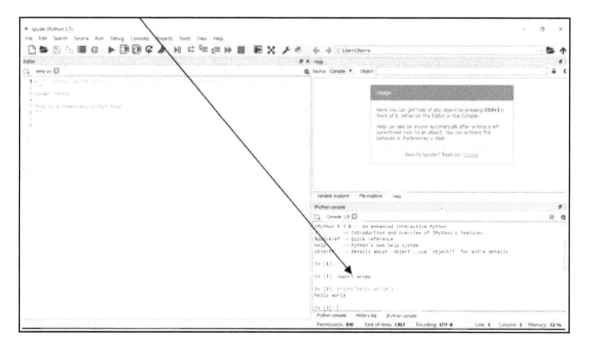

Reviewing Conda basics

There are a few commands that you should familiarize yourself with if you are going to be working with Conda and virtual environments. If you want to see all the packages that are included within your Conda environment, then you can use `conda list`:

```
Administrator: Windows PowerShell                                    —    □    ×

PS C:\coding\python> .\activateproenv.bat

C:\coding\python>call    & ("c:\Program Files\ArcGIS\Pro\bin\Python\Scripts\proenv.bat")
Deactivating environment "c:\Program Files\ArcGIS\Pro\bin\Python"...
Activating environment "c:\Program Files\ArcGIS\Pro\bin\Python\envs\arcgispro-py3"...

[arcgispro-py3] C:\coding\python>conda list
# packages in environment at c:\Program Files\ArcGIS\Pro\bin\Python\envs\arcgispro-py3:
#
DEPRECATION: The default format will switch to columns in the future. You can use --format=(legacy|columns) (or define a
 format=(legacy|columns) in your pip.conf under the [list] section) to disable this warning.
alabaster            0.7.10              py35_0     defaults
arcgispro            1.4                 0          esri
astroid              1.4.9               py35_0     defaults
babel                2.3.4               py35_0     defaults
bleach               1.5.0               py35_0     defaults
chardet              2.3.0               py35_0     defaults
colorama             0.3.7               py35_0     defaults
cycler               0.10.0              py35_0     defaults
decorator            4.0.11              py35_0     defaults
docutils             0.13.1              py35_0     defaults
entrypoints          0.2.2               py35_1     defaults
freetype             2.6.3               vc14_1     [vc14]  esri
future               0.16.0              py35_1     defaults
html5lib             0.999               py35_0     defaults
icu                  57.1                vc14_0     [vc14]  defaults
imagesize            0.7.1               py35_0     defaults
ipykernel            4.5.2               py35_0     defaults
ipython              5.3.0               py35_0     defaults
ipython-genutils     0.1.0               <pip>
ipython_genutils     0.1.0               py35_0     defaults
isort                4.2.5               py35_0     defaults
jedi                 0.9.0               py35_1     defaults
jinja2               2.9.5               py35_0     defaults
jpeg                 9b                  vc14_0     [vc14]  defaults
jsonschema           2.5.1               py35_0     defaults
jupyter-client       5.0.0               <pip>
jupyter-core         4.3.0               <pip>
```

To see information about your current Conda working environment, use `conda info`:

```
Administrator: Windows PowerShell

[arcgispro-py3] C:\Coding\Python>conda info
Current conda install:

               platform : win-64
          conda version : 4.0.11
    conda-build version : not installed
         python version : 3.5.2.final.0
       requests version : 2.11.1
       root environment : C:\Program Files\ArcGIS\Pro\bin\Python  (writable)
    default environment : C:\Program Files\ArcGIS\Pro\bin\Python\envs\arcgispro-py3
       envs directories : C:\Program Files\ArcGIS\Pro\bin\Python\envs
          package cache : C:\Program Files\ArcGIS\Pro\bin\Python\pkgs
           channel URLs : https://conda.anaconda.org/esri/win-64/
                          https://conda.anaconda.org/esri/noarch/
                          https://repo.continuum.io/pkgs/free/win-64/
                          https://repo.continuum.io/pkgs/free/noarch/
                          https://repo.continuum.io/pkgs/pro/win-64/
                          https://repo.continuum.io/pkgs/pro/noarch/
            config file : C:\Program Files\ArcGIS\Pro\bin\Python\.condarc
      is foreign system : False


[arcgispro-py3] C:\Coding\Python>
```

As you may remember, in the previous section, we installed the Spyder IDE using the `conda install` command. This will install the packages that you want to use in your environment. For example, if we want to install **Jupyter**, which is the package that allows us to use a browser-based Python environment, we can just type `conda install jupyter`:

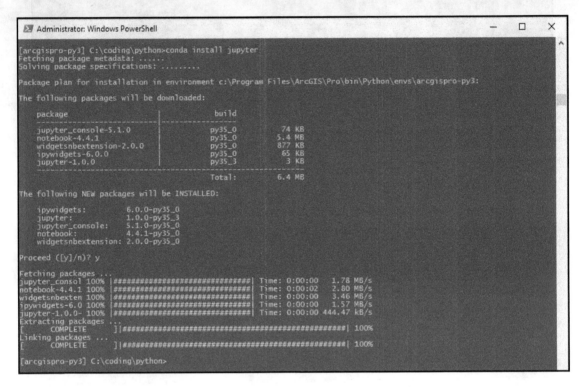

Then, if you want to run the Jupyter Notebook, then just type in `jupyter notebook`, and it should launch in your default browser:

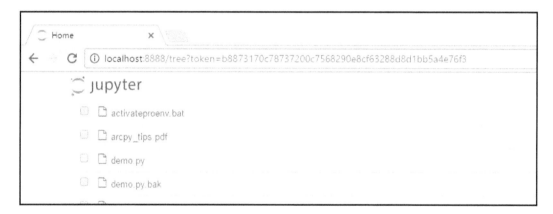

You should see a tab open in your browser (you may have to copy and paste the URL token from the command line):

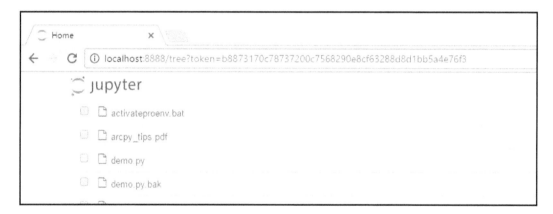

We will be using and learning more about Jupyter in the next chapter.

Summary

In this chapter, we introduced ArcGIS Pro's Python 3.5 environment and talked about the use of Conda, the Python package management system distributed with ArcGIS Pro. This is the beginning of the future of ArcGIS Python development. It is an interesting and exciting time to learn how to program in the ArcGIS environment. ESRI is undergoing a transition between their current 32-bit ArcMap applications to the modern 64-bit ArcGIS Pro app. If you intend to continue working with ArcGIS and Python, you can expect this to become the standard. In the next chapter, we will begin to learn the ArcGIS Python API environment, which is dependent on much of the same technology we learned about in this chapter.

12
ArcGIS API for Python

After reviewing some of the basics of Python and covering the foundational knowledge required to program with Python using ArcGIS Online and ArcGIS Enterprise, we can now move on to the newest release by ESRI, the ArcGIS API for Python. ESRI has been working on an easy and efficient solution to combine desktop scripting with online web mapping and infrastructure. The ArcGIS API for Python is ESRI's most modern solution to scripting for web mapping with Python. In the next chapter, we will cover the basics of the ArcGIS API for Python and learn how to configure our environment to work with the API using two versions of Python.

In this chapter, we will cover the following topics:

- Introducing the ArcGIS API for Python
- Installing Anaconda and working with Jupyter
- Introducing the four primary modules
- Providing script examples to create a web map and interact with data
- An example of administering an ArcGIS Online account

Introduction to the ArcGIS API for Python

The ArcGIS API for Python is the interface in which you can program your on-premise and cloud GIS web services using Python. In December 2016, ESRI officially released the 1.0 version of the API. It is considered a *Pythonic* implementation of an API, which means it conforms to the best practices in its design and uses the standard data structures any professional Python programmer would be familiar with. It begins to implement some of the long-held best practices that are used by traditional programmers for the GIS professional. To get started with the API, we will need to configure our machines and begin to learn a new environment for Python programming using Anaconda and Jupyter.

Installing and configuring Anaconda with Jupyter

In the previous chapter, we covered some of the basics of Python 3.5 and the Conda environment used with ArcGIS Pro. If you configured your Conda environment and feel comfortable with it, you can continue to use it. Everything we cover in this chapter will work with the Conda installation that comes with ArcGIS Pro. On the other hand, if you don't have ArcGIS Pro or would like to use a different environment and you want to use the ArcGIS API for Python, then you will have to install a product called Anaconda. In the beginning, it might be a little confusing to understand the difference between Conda and Anaconda, but Anaconda is a large platform that comes preconfigured with Python, numerous Python packages, Conda, and an IDE. Anaconda is designed by a company named Continuum Analytics. To get started, we can navigate to `https://www.continuum.i o/downloads`and install the 64-bit version of Anaconda with the Python 3.x version. When you click on the installer, you should see the installation wizard shown as follows:

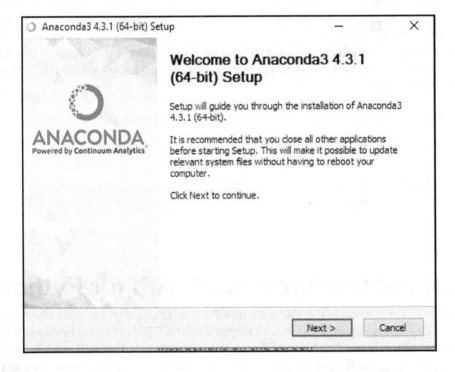

The next window gives you the option to choose between **Just Me** or **All Users**; it is recommended that you choose **Just Me**. This way, you do not need administrative privileges to use and install additional packages:

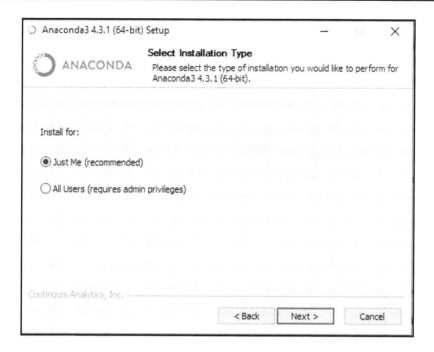

Next, you can choose the default option:

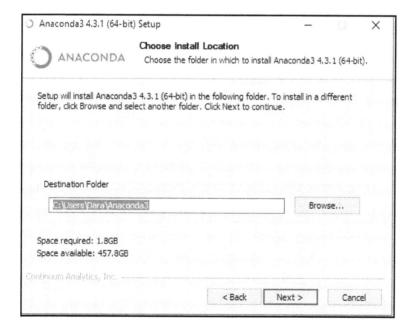

The next screen is very important. You want to make sure you uncheck both options. If you work with ArcGIS Desktop and Python 2.7, this will break your configuration. You want to make sure you *do not* add this to your **PATH** environment variable or make it your default Python:

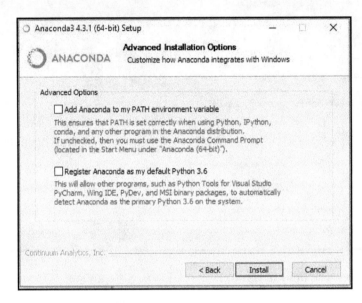

Then, click on **Install**, and it should get installed with no problem:

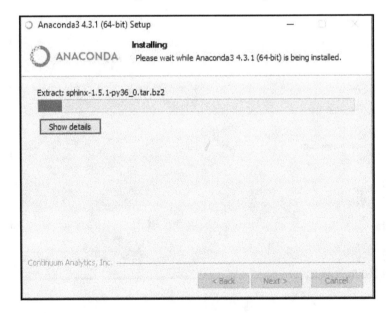

Once Anaconda has finished installing, you should be able to go to your start menu and see a folder for **Anaconda** located there:

If you open that folder, then you should see various packages. We will open **Anaconda Prompt** and it will look like this:

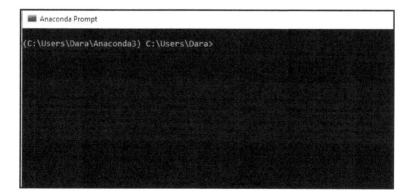

If you have permission issues with the installation, try to run **Anaconda Prompt** explicitly as an administrator by right-clicking and choosing run as administrator.

Install the ArcGIS Python API

Next, you will be installing the ArcGIS API for Python; type in the following:

```
conda install -c esri arcgis
```

Then, follow any onscreen prompts:

Once this is complete, you should see an `arcgis-python-api` folder located in your `C:\users\username` folder:

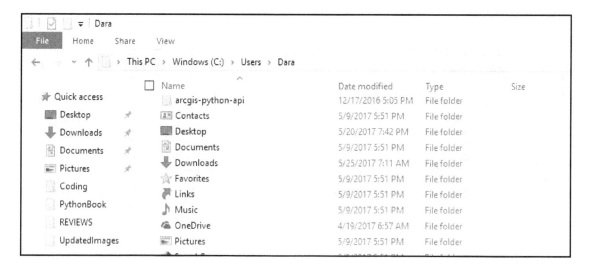

At this point, you have Anaconda installed, which comes with Jupyter and other packages. If Jupyter did not get installed as expected, you can always install it with the `conda install jupyter` command. Next, we will begin the process to set up Jupyter.

Creating a Jupyter Notebook

Jupyter Notebook is an open source browser-based application that allows you to execute Python code and thereby easily share it, along with any visualizations you might create with the code. If you have ever heard of *iPython* Notebook, it is very similar to that. In this section, I will discuss how to create a notebook and some of the basic functions you will need to know in order to work with it.

To get started, let's launch that Anaconda Prompt again, from your start menu. Then, type the `jupyter notebook` command, and if everything is installed correctly, it should launch in your default browser automatically:

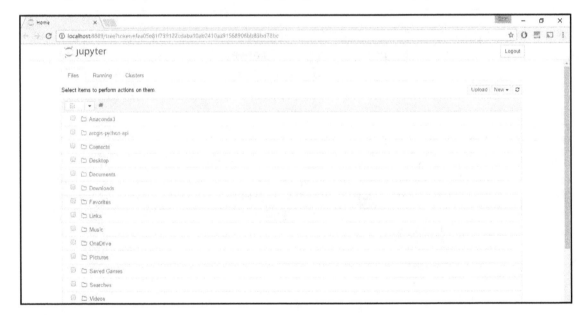

When opened, you should see a browser page open that looks like this:

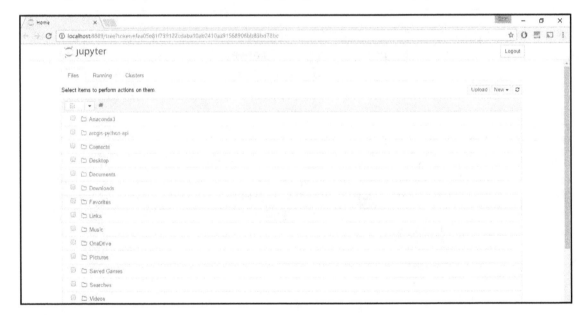

You should notice that this directory is in the same location you were in when you launched the notebook from Command Prompt. In my case, this is my `C:\Users\Dara` directory. To begin coding with Jupyter, we need to create a new notebook. In the top-right corner, you should see the **New** drop-down option:

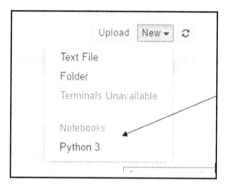

Click on the **Python** 3 Notebook and a new tab will open.

Click on **Untitled** at the top:

You will be prompted to rename your notebook:

Now we begin to experiment with running code in Jupyter. In the first line, enter this:

```
print("Hello Juptyer")
```

Then, in order to execute code, there are a couple of options you have. First, you can click on the **run cell, select below** tool on the toolbar at the top of the page:

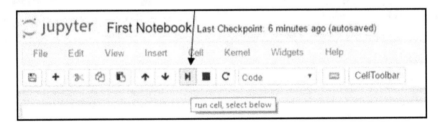

Or, as I prefer, you can use the keyboard shortcut *Ctrl + Enter* to execute the cell. Then, in order to insert a new cell, you can enter **b**, as shown here:

If you look in our home directory, you should see the file we started working on called `First Notebook.ipynb`; this is the same notebook we are working on that can be shared with other users.

Another useful tip when working with Jupyter is the use of code completion. If we import the math module and then enter **math.** + *Tab*, you should see the list of available methods in the math module:

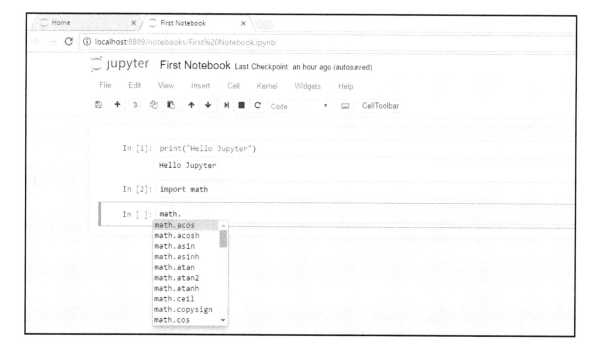

This will be useful when we begin to work with the Python API. Some other useful keyboard shortcuts include the following:

- **d + d (tapping "d" twice) :** This will delete the highlighted cell
- **shift + tab :** This will show you the docstring of the object your cursor is on:

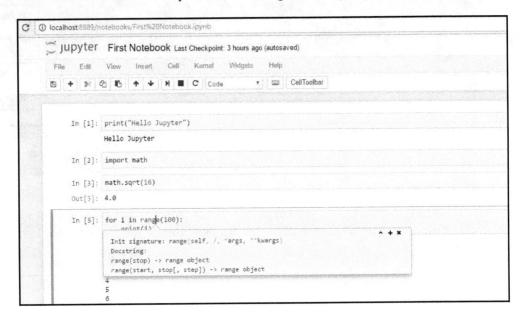

- **shift + m :** This merges multiple cells as shown:

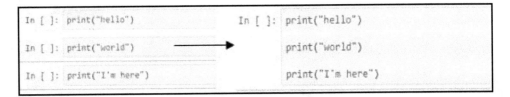

To see more shortcut key commands, click on the command pallet tool:

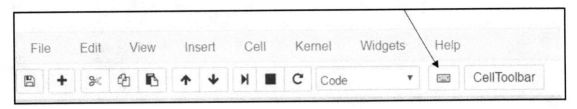

Another great aspect of the Jupyter Notebook is that when you run code and there is an error, you can dynamically fix that code and rerun the same cell. For example, we loop over a range of 100 and print each number, but the first time we run it, we forget to put parenthesis in the print function:

```
In [1]:  print("Hello Jupyter")

         Hello Jupyter

In [2]:  import math

In [3]:  math.sqrt(16)

Out[3]:  4.0

In [4]:  for i in range(100):
             print i

           File "<ipython-input-4-f90dfd5040b1>", line 2
             print i
                   ^
         SyntaxError: Missing parentheses in call to 'print'
```

We can just adjust the code and rerun that cell:

```
In [1]:  print("Hello Jupyter")

         Hello Jupyter

In [2]:  import math

In [3]:  math.sqrt(16)

Out[3]:  4.0

In [5]:  for i in range(100):
             print(i)

         0
         1
         2
         3
         4
         5
         6
         7
```

In Jupyter, you can also pass commands similar to power shell. For example, if you want to see your working directory or the contents of your working directory, you can execute **ls** or **pwd**.

You can do that using **ls**:

```
In [6]: ls
         Volume in drive C is Windows
         Volume Serial Number is 1AA5-03DB

         Directory of C:\Users\Dara

04/08/2017   04:57 PM    <DIR>          .
04/08/2017   04:57 PM    <DIR>          ..
12/17/2016   09:10 PM    <DIR>          .anaconda
08/19/2016   10:34 PM    <DIR>          .atom
12/18/2016   04:46 PM              412 .bash_history
12/17/2016   03:46 PM    <DIR>          .conda
01/25/2017   08:22 PM              157 .gitconfig
02/13/2017   06:15 PM    <DIR>          .idlerc
```

You can also do that using **pwd**:

```
In [7]: pwd
Out[7]: 'C:\\Users\\Dara'
```

Starting the ArcGIS API for Python

Now that you are a little more familiar with Jupyter and using the Notebook environment, we can begin to explore how to use the ArcGIS API for Python. The API is designed to follow the standard practices within the Python community and is implemented as a modular API. What does this mean exactly? Every aspect of the API is accessed by calling a module in the Python API. This is similar to the ArcPy standard. For example, calling `arcpy.mapping` accesses a method within a module, which is referred to as modular.

In the language used to describe the API, the generic phrase "your GIS" is referred to as working with any part of the ESRI web platform, that is, ArcGIS Online or ArcGIS Enterprise, not to be confused with `gis` or `GIS`. The most important and most used module in the ArcGIS API for Python is the `gis` module. The `gis` module allows you to manage groups, users, and content in your GIS. To invoke the `gis` module as an anonymous user, you use the following code:

```
from arcgis.gis import GIS
gis = GIS()
sf_map = gis.map("San Francisco")
sf_map
```

In the preceding example, we import the `GIS` object from the `gis` module in the first line and then we create an instance of the `GIS` object and assign it to the `gis` variable in the second line. The fourth line of code calls the `sf_map` object and has the notebook visualize the object shown, as follows:

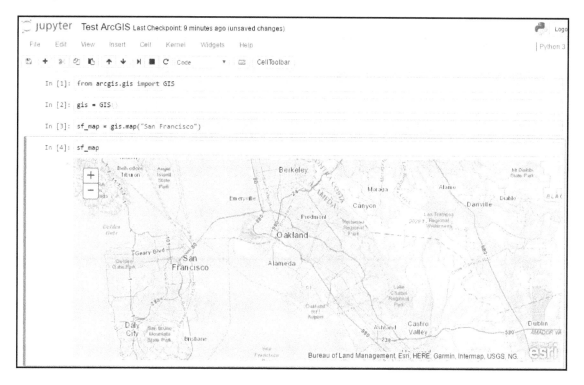

To invoke a `GIS` object from an organizational account, such as your ArcGIS Online developer account, you need to pass in the URL of your account and the login credentials, as shown:

```
from arcgis.gis import GIS
gis = GIS("https://www.arcgis.com", "username", "password")
sf_map = gis.map("San Francisco")
sf_map
```

The preceding code is implemented as follows:

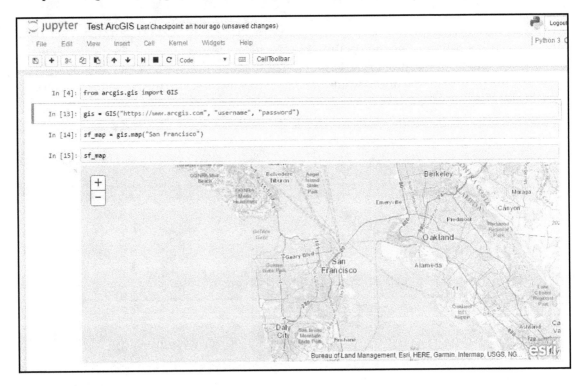

Adding an item to a web map

Once you have passed the credentials of your GIS or organizational account to the `GIS` object, you can access the items in your account and add them to a map. First, we can visualize all the items within your organization. To do this, we will import the `IPython.display` library and then we will use the `content.search` method to find all items with `'SF'` in the description. After we have found all the items that have `'SF'` in them, we can loop over each item and display them:

```python
from IPython.display import display
items = gis.content.search('SF')
for item in items:
    display(item)
```

The preceding code is implemented as follows:

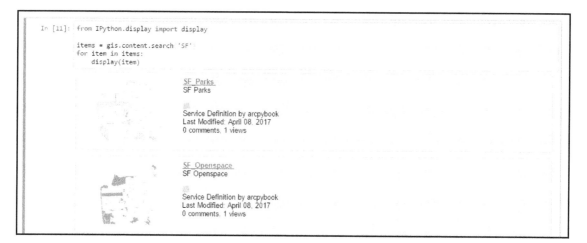

Next, we can use a list comprehension to find `SF_BusStops` and add it to our `sf_map` object we created earlier:

```python
add_item = [bus_item for bus_item in items if bus_item.title ==
"SF_BusStops"]
sf_map.add_layer(add_item[0])
```

Following screenshot represents the preceding code in Jupyter:

```
In [12]: add_item = [bus_item for bus_item in items if bus_item.title == "SF_BusStops"]
         sf_map.add_layer(add_item[0])
```

If you scroll up in Jupyter back to the top where we created our sf_map object, you should see SF_BusStops added to the map like this:

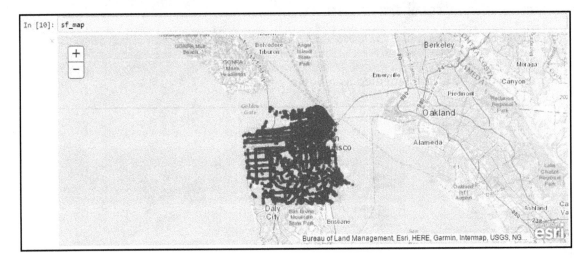

Importing a CSV with pandas

As a GIS professional, you encounter many data types that you have to manage. Much of the data you will manage is commonly in a **comma-separated value** (**CSV**) format. Later, I will show you how to add a CSV as an item to your ArcGIS Online organization. In addition, if that CSV has a Lat/Long or X/Y, you can also add it as a spatial feature layer to you GIS.

To work with CSVs, we will be using an open source Python library called **pandas**. First, we import pandas and call it `pd` and then we build a data frame using the `read_csv` function from pandas. A data frame is a structured way of viewing data; you can think of it as a spreadsheet or an SQL table. In the next figure, you can see what a data frame looks like when we build it from our `SFPF_2016.csv`. For the purposes of this exercise, I have reduced `SFPD_2016.csv` from its original size of 150,000 records down to 5,000:

```
import pandas as pd
dataFrame=
pd.read_csv("C:\PythonBook\ch12_ArcGIS_PythonAPI\Ch12_Data\SFPD_2016.csv",
encoding = "ISO-8859-1")
```

This is how the data frame looks:

Next, we can use the `.describe()` function in the `PdDistrict` field to see summary information about that data field:

```
dataFrame['PdDistrict'].describe()
```

The preceding code is implemented as follows:

Then, we can use the `matplotlib` Python library to create a simple histogram plot on our `PdDistrict` field. After we import the library, we access the `PdDistrict` field in the data frame and then use the `.value_counts()` function to get a count of each unique record. Finally, we pass `.plot(kind='bar')` to return a visualization of our data. Feel free to experiment with the `kind` option. Not only can you use `bar`, but you can also use box, pie, line, and area:

```
import matplotlib.pyplot as
py_plotdataFrame['PdDistrict'].value_counts().plot(kind='bar')
```

The preceding code is implemented as follows:

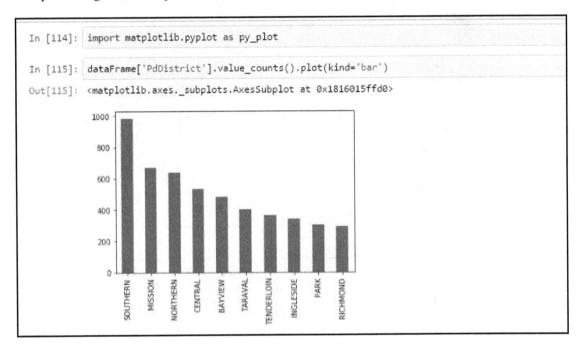

After we are done visualizing our data fields, we can actually add this CSV as an item to our ArcGIS Online account. We can pass the CSV file to the `pd_data` variable and then build the properties for this object. Under the properties, you will see that we have title, description, and the tags that will be assigned to the item in ArcGIS Online. Then, we can use a thumbnail that will appear next to the item on ArcGIS Online. Finally, we will use the `gis.content.add()` function to actually add the data to ArcGIS Online and provide the property information, along with the thumbnail:

```
pd_data = r"C:\PythonBook\ch12_ArcGIS_PythonAPI\Ch12_Data\SFPD_2016.csv"
pd_data_properties = {'title': 'SFPD calls for the year 2016',
```

```
'description': 'All the SFPD calls for the year 2016','tags': 'SF PD,
calls, csv' }
thumbnail_pic = r"C:\PythonBook\ch12_ArcGIS_PythonAPI\Ch12_Data\SF_PD.PNG"
sf_pd_item = gis.content.add(item_properties=pd_data_properties,
data=pd_data, thumbnail = thumbnail_pic)
sf_pd_item
```

The preceding code is implemented as follows:

```
In [8]:  pd_data = r"C:\PythonBook\ch12_ArcGIS_PythonAPI\Ch12_Data\SFPD_2016.csv"
         pd_data_properties = {'title': 'SFPD calls for the year 2016',
                               'description': 'All the SFPD calls for the year 2016',
                               'tags': 'SF PD, calls, csv' }

         thumbnail_pic = r"C:\PythonBook\ch12_ArcGIS_PythonAPI\Ch12_Data\SF_PD.PNG"

         sf_pd_item = gis.content.add(item_properties=pd_data_properties, data=pd_data,
                               thumbnail = thumbnail_pic)

         sf_pd_item
```

```
Out[8]:
```

SFPD calls for the year 2016

CSV by arcpybook
Last Modified: April 20, 2017
0 comments, 0 views

After we have created the item on ArcGIS Online, we can publish it as a hosted feature layer. If you remember, I said earlier that if our CSV had coordinate information, we could make it a spatial layer. To publish the item as a feature layer, we use the `.publish()` method available in the `gis` module:

```
sf_pd_feature_layer = sf_pd_item.publish()
sf_pd_feature_layer
```

The preceding code is implemented as follows:

```
In [9]: sf_pd_feature_layer = sf_pd_item.publish()
        sf_pd_feature_layer
```

```
Out[9]:
```

SFPD calls for the year 2016

Feature Layer Collection by arcpybook
Last Modified: April 20, 2017
0 comments, 0 views

Once the item has been published as a feature layer, we can visualize the data on a map. First, we create a variable called `sf_crimemap` and then we can even use an intersection in San Francisco as the location. We can also set the zoom level to 17. Then, we use the `.add_layer()` function to add the `sf_pd_feature_layer` to `sf_crimemap`:

```
sf_crimemap = gis.map("Divisadero and Haight,San Francisco", zoomlevel=17)
sf_crimemap.add_layer(sf_pd_feature_layer)
sf_crimemap.height = '950px'
sf_crimemap
```

The preceding code is implemented as follows:

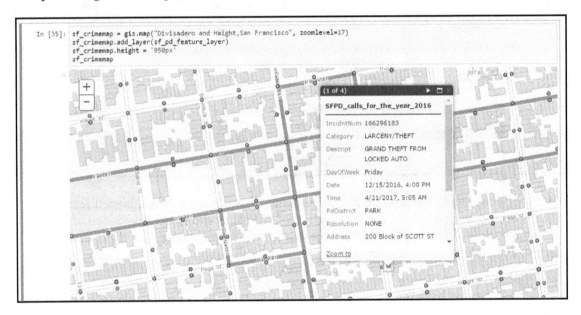

Summary

In this chapter, we covered the use of the ArcGIS API for Python. We introduced you to Anaconda and a more advanced function of Jupyter. These pieces of infrastructure assist the GIS professional as the ArcGIS platform undergoes a migration from Python 2.7 to 3.6. It enables you to easily work with different Python environments. In addition to learning about new technology in this chapter, we also reviewed the methods to add items to our ArcGIS Online account.

Index

A

Anaconda
  configuring, with Jupyter 226
  installing, with Jupyter 226
Application Programming Interface (API) 9
ARC Macro Language (AML) 34
ArcGIS API
  Anaconda, configuring with Jupyter 226
  Anaconda, installing with Jupyter 226
  ArcGIS Python API, installing 230
  for Python 225
  starting, for Python 238
ArcGIS Online (AGOL)
  about 133
  developer account 152
  interface, exploring 136
  MXD, publishing from 142
  signing up, for account 134
  URL, for signing up 134, 152
ArcGIS Online administration 184
ArcGIS Online REST Services
  about 155
  exploring 156
  implementing 172
  reference link 159, 164
ArcGIS Online tokens
  URL, building 170
ArcGIS Pro Python window 205, 206, 207, 208,
    209
ArcGIS Pro
  about 202, 212
  configuring 202, 203, 204
  installing 202, 203, 204
ArcGIS REST API
  references 161
arcgispro-py3 212

ArcPro
  Python 2.7 210, 211
  Python 3.5 210, 211
ArcPy Array objects 72
ArcPy geometry object classes
  about 70
  ArcPy Array objects 72
  ArcPy Point objects 71
  ArcPy PointGeometry objects 79
  ArcPy Polygon objects 75
  ArcPy Polyline objects 73
  bus stop analysis, rewriting 80
ArcPy Point objects
  about 71
  reference link 72
ArcPy PointGeometry objects 79
ArcPy Polygon objects
  about 75
  AsShape method 78
  generic geometry object 79
  methods 77
  reference link 75
ArcPy Polyline objects
  about 73
  reference link 74
ArcPy tools
  about 44
  data, accessing cursor used 46
  Intersect tool 45
arcpy.AddMessage
  used, for displaying script messages 88
arcpy.mapping module
  data source 107
  definition queries 107
  description 107
  layer object methods 106
  layer object properties 106

name 107
visibility 107
ArcPy
  layer sources, inspecting 108
  layer sources, replacing 108
  map document elements, interacting with 104
  map document production, automation 110
  using, with map documents 104
ArcREST
  about 180
  installing 180
  package structure 181
  security handler 183
  URL 180
ArcToolbox 12, 39
AsShape method
  reference link 78
Atom 28
attribute field interactions 60
automatically generated script 38
Avenue 34

B

built-in functions 21
bus stop analysis
  adding to 82
  rewriting 80
Bus Stop feature class
  adding 94

C

cartographic 201
Census Block feature class
  adding 95
Census Block field
  adding 96
comma-separated value (CSV)
  about 97, 242
  importing, with pandas 242, 243, 244, 246
comments, Python 13
Conda
  about 201, 212
  basics, reviewing 220, 221, 222, 223
  standalone scripts, executing 213, 214, 216, 217, 218, 219

Continuum Analytics
  reference link 226
Copy Features tool 70
CSV module 21
cursors
  updating 61

D

data access module
  about 57
  attribute field interactions 60
  cursors, updating 61
  Insert cursor, using 63
  point location, adjusting 62
  polygon geometry, inserting 66
  polyline geometry, inserting 65
  row, deleting Update cursor used 63
  search cursors 58
  shape field, updating 62
data containers 15
data types
  about 14
  adding 94
  data containers 15
  dictionaries 17
  floats 15
  integers 14
  lists 16
  strings 14
  tuples 17
  zero-based indexing 15
data
  accessing, cursor used 46
  exceptions 48
  files, overwriting 48
  tracebacks 48
dataList 60
definition queries 107
dictionaries 17
dynamic components
  adding, to script 89
dynamic parameters
  adding, to script 85
  dynamic components, adding to script 88
  passed parameters, accessing 86

script messages, displaying arcpy.AddMessage
   used 88

E

elif statement, Python 12
else statements, Python 12
EPSG coding system 75

F

feature set 162
feature set method
   about 163
   load method 163
   save method 163
file paths
   in Python 40
final script
   inspecting 99
   script tool, executing 101
floats
   about 15
   reference link 15
folder structure, Python
   about 28
   modules, location 28
   sys module, used for adding module 30
   sys.path.append method 30
   third-party module, installing 29
for loops, Python 12
functions, Python 19

G

geodatabase
   reference link 34
GeoJSON 78
glue language 9

I

IDLE 27
IDLE (Python GUI) 33
if statement, Python 12
in memory 186
indentation, Python 19
Insert cursor

using 63
integers 14
Integrated Development Environments (IDEs)
   about 7, 27
   Atom 28
   conclusion 28
   IDLE 27
   Notepad++ 28
   PythonWin 27
   Sublime Text 28
interface, AGOL
   exploring 136
   Groups tab 141
   Map tab 142
   My Content tab 137
   My Organization tab 137
   Scene tab 142
interpreted language 8
Intersect tool
   about 45
   CSV module, adding to script 45
   script, adjusting 45
item
   adding, to web map 241, 242

J

JavaScript Object Notation (JSON) 57, 78, 161, 193
join 117
Jupyter Notebook
   creating 231, 233, 234, 236, 237, 238
Jupyter
   about 222
   Anaconda, configuring 226
   Anaconda, installing 226

K

Keyhole Markup Language (KML) 57
keywords, Python 19

L

layer sources
   ListBrokenDataSources method 108
   PDF, exporting from MXD 110
ListBrokenDataSources method

broken links, fixing 108
links, fixing of individual layers 109
lists 16

M

map document elements
  arcpy.mapping module, using to control layer
    objects 106
  data frames 104
  pan methods 105
  zoom methods 105
map document production
  automation 110
  Dynamic maps 119
  variables 113
map documents
  ArcPy, using 104
model
  automatically generated script 38
  exporting 38
  Python, file paths 40
ModelBuilder
  about 33, 34
  analysis results, tallying 37
  Buffer tools, modeling 35
  creating 34
  exporting, to Python 34
  Intersect tool, adding 36
  Select, modeling 35
MXD
  layers, publishing 144
  layers, styling 143
  publishing from 142
  Service Editor 148
  Share As menu 147
  updates 151
My Content tab
  Add Item option 138
  Create tab 140
  files, features 139
  services, features 139

N

namespaces, Python 20
Notepad++ 28

O

Open Geospatial Consortium (OGC) 78
OS (operating system) module 20
output spreadsheet
  adding 97
  bus stop fields, adding 98
  field names, adding 97
  SQL Statement, adding 98

P

pandas
  about 243
  CSV, importing 242, 243, 244, 246
parameters
  defining 92
  labeling 92
  of Python functions 54
passed parameters
  accessing 86
point location
  adjusting 62
polygon geometry
  inserting 66
polyline geometry
  inserting 65
Python 2.7
  with ArcPro 210, 211
Python 3.5
  with ArcPro 210, 211
Python functions
  about 52
  createCSV function 56
  creating 53
  technical definition 52
  used, for replacing repetitive code 55
  with parameters 54
  XLS, creating XLWT used 56
Python interpreter
  about 22
  location 23
  location, known by machine 24
  reference link 39
  using 23
Python modules

about 20
built-in functions 21
CSV module 21
location 28
OS (operating system) module 20
standard library modules 21
sys (Python system) module 20
XLRD module 21
XLWT module 21
Python Package Index (PyPI) 29
Python programming
basics 9
comments 13
elif statement 12
else statements 12
for loops 12
if statement 12
statements, importing 10
variables 11
while statements 13
Python Prompt 206
Python script
about 22
executing 26
Python
ArcGIS API 225
ArcGIS API, starting 238
concepts 18
functions 19
glue language 9
indentation 19
interpreted language 8
keyword 19
namespaces 20
script, executing 22
standard (built-in) library 8
URL, for downloading 23
used, as programming language 8
wrapper modules 9
PythonWin 27

Q

query hosted feature services
about 185
domain, adding to fields 191

feature class, appending to feature service 194
features, querying 186
features, saving as feature class 186
Field, adding to Feature Service 188
records, updating 196

R

Representational State Transfer (REST) 155
row
deleting, Update cursor used 63

S

script tool
Bus Stop feature class, adding 94
Census Block feature class, adding 95
Census Block field, adding 96
creating 90
data types, adding 94
executing 101
output spreadsheet, adding 97
parameters, defining 92
parameters, labeling 92
script
about 49
adjusting 38
dynamic components, adding 88
dynamic parameters, adding 85
messages, displaying arcpy.AddMessage used 88
search cursors 58
Service Editor
Analyze 150
Item Description option 149
shape field
updating 62
standalone scripts
executing, with Conda 213, 214, 216, 217, 218, 219
standard (built-in) library, Python 8
standard library modules
about 21
string manipulation
adding 41
formatting method 1 42
formatting method 2 43

strings 14
Sublime text 28
sys (Python system) module 20
sys module
  used, for adding module 30
sys.path.append method 30

T

third-party module
  installing 29
Transcript 206
tuples 17

U

Update cursor
  used, for deleting row 63
URL parameters
  about 159
  format 159
US Census
  URL 34

V

variables
  about 11
  adjusted map, exporting to PDF 118
  buffer, generating from bus stops feature class

  115
  bus stop buffer, intersecting 115
  census blocks, intersecting 115
  data frames 114
  dynamic definition query, formatting 116
  layers, accessing 114
  layout elements 114
  layout elements, updating 118
  map document, connection to 113

W

web map
  item, adding 241, 242
Well-Known Binary (WKB) 57
Well-Known Text (WKT) 57
while statements, Python 13
wrapper modules 9

X

XLRD module 21
XLRD
  reference link 56
XLWT module 21
XLWT
  reference link 56

Z

zero-based indexing 15